THE SPACE AGE GENERATION

THE SPACE AGE GENERATION

Lives and Lessons from the Golden Age of Solar System Exploration

EDITED BY

WILLIAM SHEEHAN AND KLAUS BRASCH

THE UNIVERSITY OF
ARIZONA PRESS

TUCSON

The University of Arizona Press
www.uapress.arizona.edu

We respectfully acknowledge the University of Arizona is on the land and territories of Indigenous peoples. Today, Arizona is home to twenty-two federally recognized tribes, with Tucson being home to the O'odham and the Yaqui. Committed to diversity and inclusion, the University strives to build sustainable relationships with sovereign Native Nations and Indigenous communities through education offerings, partnerships, and community service.

ISBN-13: 978-0-8165-5104-0 (hardcover)
ISBN-13: 978-0-8165-5105-7 (ebook)

Cover design by Leigh McDonald
Cover photos: (*front*) The vintage image used on the cover is characteristic of an era when young boys were largely the audience targeted for astronomy and space. Artist unknown. (*back*) Heidi Hammel fills the dewar holding a CCD camera that sits at the bottom of the tube at the Cassegrain focus of the 88-inch telescope at Mauna Kea, Hawai'i. Photo by Dale P. Cruikshank.
Designed and typeset by Sara Thaxton in 10/14 Warnock Pro with Iva WF, Payson WF, and Helvetica Neue LT Std

Library of Congress Cataloging-in-Publication Data
Names: Sheehan, William, 1954– editor. | Brasch, Klaus, 1940– editor.
Title: The space age generation : lives and lessons from the golden age of solar system exploration / edited by William Sheehan and Klaus Brasch.
Description: Tucson : University of Arizona Press, 2024. | Includes bibliographical references and index.
Identifiers: LCCN 2023031896 (print) | LCCN 2023031897 (ebook) | ISBN 9780816551040 (hardcover) | ISBN 9780816551057 (ebook)
Subjects: LCSH: Planetary science—History—20th century. | Outer space—Exploration—History—20th century.
Classification: LCC QB501 .T44 2024 (print) | LCC QB501 (ebook) | DDC 523.4092/2—dc23/eng/20230927
LC record available at https://lccn.loc.gov/2023031896
LC ebook record available at https://lccn.loc.gov/2023031897

Printed in the United States of America
♾ This paper meets the requirements of ANSI/NISO Z39.48-1992 (Permanence of Paper).

*Dedicated to future generations
of solar system explorers*

CONTENTS

PREFACE

This book grew out of a November 2021 article, "A Golden Age for Amateur Astronomy," that we published in *Sky & Telescope* on the eightieth anniversary of the magazine's first publication. The result of a merger of unsuccessful magazines *The Sky* and *The Telescope*, the new offspring survived World War II and the first postwar years. Only in the fifties and sixties, however, did it really come into its own. *Sky & Telescope* was successful for many reasons, including the increase in discretionary income during the postwar economic boom, which those inclined—after spending on new cars or TVs or dishwashers and washing machines—could spend on good telescopes. Even more significant, however, was that those were the years in which the space age got under way—with the first high-altitude rocket tests, artificial satellites, and probes to the Moon and planets.

Though we didn't fully realize it at the time, we now look back to that era (as suggested in the title of our article) as something of a golden era for amateur astronomy. We found out, through the rather unprecedented response to our article by its increasingly aging readership, that it was a golden age for many others who grew up in that era as well, of whom not a few went on from humble beginnings building their own telescopes and making their own observations (it was "useful work," we were assured) to become leading professional scientists involved in establishing the modern multidisciplinary science of the solar system. Of course, nowadays, with the range of sophisticated images taken even by amateurs with skill, leisure, and means, using charge-coupled devices (CCDs), as well as the onslaught of data being sent back by spacecraft and readily accessible on the internet, the activities that absorbed us "in the good old days" are likely to seem rather tame. We certainly didn't know very much, and a single enthusiastic author, like England's Patrick Moore, could cover the waterfront with a series of books that included nary an equation and aimed its pitch perfectly to the level of any intelligent twelve-year-old. That so little was known, however, meant that the whole solar system was a realm of mystery, and it was the mystery that bestowed a peculiar piquancy to the whole enterprise—an enterprise in which we engaged in sketching lunar craters on lunar limbs or capturing visually fine details, such as canals on the surface of Mars or wisps in the zones of Jupiter, contributing our widow's mite to the grand edifice of science. Indeed, until the nineties, the human eye remained supreme as a detector of planetary detail, only at long last to be cast into the shade through digital technology.

Every generation has its nostalgia and tends to think of itself as special. But those of us who belonged to the generation growing up during the space age were fortunate to live through a unique era, in which the Moon and planets were studied for the first time by spacecraft at close range. Though our knowledge now continues to increase by leaps and bounds, humans will never be so innocent again as we were then, never perhaps believe, with as much naïveté, that we could solve our problems on Earth and step out boldly and confidently across space from world to world. Ours was in a way a virginal time. We can never experience the "first time" twice.

The response to our *Sky & Telescope* article, which included some of those who agreed to contribute essays to this volume, shows that our feelings were widely reciprocated. The University of Arizona Press was also enthusiastic about the project. Thus we came to record some of the memories of a period that is now rapidly fading away, and that will soon be lost and indecipherable as time—or Lethe—irresistibly moves forward and plows them under.

For future generations interested in the history of humanity's explorations of the solar system, we hope that this book may be useful in conveying some sense of what it was like for us to have lived through all that, even as they busily set about creating their own memories.

William Sheehan
Klaus Brasch
Flagstaff, Arizona, USA
February 17, 2023

ACKNOWLEDGMENTS

The editors would like to thank Gary Seronik of *Sky & Telescope* magazine for commissioning the article "A Golden Age for Amateur Astronomy" (November 2021), which was well received and led to brainstorming the present project. Allyson Carter of the University of Arizona Press saw the possibilities in an autobiographical book by planetary scientists who had entered and reached midcareer during the exciting if tumultuous era of the fifties, sixties, and seventies. Three anonymous reviewers contributed valuable feedback, which led to significant improvements. A book such as this, with so many different authors (and styles) can be very challenging to edit; however, copyeditor Melanie Mallon did an outstanding job of sorting out problems, while Leigh McDonald produced cover art that managed to both evoke the era and show sensitivity to readers who realize how backward things were, in some respect, in the "golden age." Each of the authors would like to thank spouses, children, colleagues, and others who contributed in ways too numerous to name, which, in an era when planetary science increasingly involves enormous teams of dedicated and skilled individuals, consists of tens of thousands of women and men—engineers, scientists, managers, technicians, administrators, students, and their support staff at universities, government labs, aerospace companies, observatories, and the major engineering and research centers run by NASA, ESA, and other space agencies. There isn't room to name them all, alas—but they know who they are!

THE SPACE AGE GENERATION

Introduction

WILLIAM SHEEHAN

The naïve idea that we knew a great deal about the solar system and that there would not be many surprises in store in future years was certainly one of the first casualties of the unmanned planetary program. It used to be that a planetary astronomer was entirely safe in the sense that he could make any speculation whatever about a planet or life thereon and could rarely be proved wrong. . . .

Clearly the best time to be alive is when you start out wondering and end up knowing. There is only one generation in the whole history of mankind in that position. Us.

—Carl Sagan, "A Very Special Time"

A Golden Age

It's not always clear that one has lived in a golden age until after the fact.

In retrospect, the period from the early fifties until the late eighties was a one-off, a golden age of planetary science. Those like us who lived through it were fortunate in belonging to the generation that was the first to explore the solar system and thereby experienced what can never be experienced again. In our childhood the planets were "distant and indistinct discs moving through the night sky, and . . . in old age, . . . places, diverse new worlds in the course of exploration."[1]

In the fifties, though full of hope, we actually knew very little. Now we know a great deal but—perhaps—are not so full of hope. In the fifties, the far side of the Moon was terra incognita, and speculation as to what might be found there was rife. The surface of Venus, cloaked under perpetually overcast skies, might be steaming jungles like those on Earth during the

---

1. Carl Sagan, *The Cosmic Connection* (New York: Doubleday, 1973), 69.

Carboniferous period. Mercury was believed to rotate in the same period as it revolved around the Sun, and so it was more or less half-baked and half-frozen—except, perhaps, in the "twilit" zone, which alternately enjoyed day and night and where life might have gotten a foothold. Mars of course was more evocative than any of the others. Percival Lowell's whims of intelligent beings and canals to pump water from the polar ice caps were still remembered, and though they were no longer viewed as likely, it seemed possible, even probable, that lower life forms, like lichens, might exist on the planet. Jupiter's Great Red Spot was possibly a large solid body floating in the planet's atmosphere like an egg in a solution of salt and water. Saturn's main rings—A, B, and C—were well defined, but the finer structure sometimes glimpsed through large telescopes in excellent seeing conditions was largely unknown, as were the forces controlling that structure. Uranus, Neptune, and Pluto were virtually inscrutable, as were the satellites of all the planets except Earth and the asteroids. The Kuiper Belt and Oort Cloud were mere theoretical speculations. Theories of the origin of the solar system and of the Earth-Moon system were primitive. Even the origin of lunar craters was hotly debated, as it had been for centuries, with keen adherents to both the meteoritic and the volcanic schools. Whether the solar system we knew was rare or commonplace in the galaxy was unclear, and we had no firm knowledge, one way or the other, of extra-solar planets. Also unknown was whether life might be rare or commonplace, though as a matter of mere statistics (with an estimated one hundred billion suns in the galaxy), it appeared exceedingly unlikely that ours was the only technologically sophisticated civilization. UFOs were all the rage, and at least a few professional astronomers believed that representatives of the planets of other stars (and perhaps even Venus or Mars) might have visited (or be visiting) our planet. The first SETI (search for extraterrestrial intelligence) programs got under way in 1960, dedicated to picking up extraterrestrial communications with radio telescopes, and though if successful, we could undoubtedly learn a great deal from civilizations more advanced than ours, the prospect of disclosing our whereabouts was not entirely without danger.

The fifties—and on into the sixties and beyond—was certainly a golden age for young people interested in science. There was unprecedented support for science education, and funding for scientific research, especially at the new space agency, NASA, shot upward. Blending astronomy, geology, physics, chemistry and biology, the new discipline of planetary science emerged to interpret the enormous amounts of data spacecraft were returning from other worlds, and within planetary science, subspecialities became more and more complex and particularized. Before long, no one person could possibly comprehend the big picture.

It was a golden age, and yet, living through it, it did not always seem like one. The period that saw the culmination of some of our oldest dreams, in which we ventured beyond the Earth, "the cradle of humanity," in Russian space visionary Konstantin Tsiolkovsky's words, came hard on the heels of a singularly horrific period of human history, which included the hideous stalemate of trench warfare in World War I; Stalin's collectivization of farming, resulting in the starvation of millions; the gulags; the Nanjing Massacre; Hitler's war; the Holocaust; and atomic bombs dropped on Hiroshima and Nagasaki. Ironically, the drive for more deadly weapons of warfare (modified ballistic missiles) became the very means that would allow voyages to the Moon and planets. It was the best and the worst of times, no different from any other except possibly more extreme.

Though the space age had many antecedents, it is usually said to have begun with the launch of Sputnik (i.e., "satellite") on October 4, 1957. Early achievements include the discovery of Van Allen radiation belts by the U.S. satellite Explorer 1 in 1958 and the three Soviet Lunas of 1959, which were, respectively, the first spacecraft to reach the vicinity of the Moon, the first human object to reach its surface, and the first spacecraft to photograph the hitherto unseen far side of the Moon. The sixties saw the first weather satellite (Tiros), communication satellites (e.g., Telstar 1), men and women in Earth orbit, more probes to the Moon as well as to Venus and Mars, and finally, in President John F. Kennedy's words, "landing a man on the Moon and returning him safely to the Earth."

The backdrop to all this was the Cold War, in which two quasi empires, the democratic and capitalist United States and the autocratic and communist Soviet Union, struggled for global dominance. Politically, space was vital not so much in itself than in the prestige it offered whoever achieved mastery of it first. Each launch was a chance to demonstrate the awesome power of rocket systems to deliver payloads into space, whether these were probes sent to explore the Moon and planets or nuclear warheads intended for nothing less than the destruction of the Earth. Regardless of these mixed motives, the result was to be a golden age of space exploration, in which humanity first extended its reach into the solar system.

The authors of the present collection of essays are among those who lived during that remarkable era and witnessed, or directly contributed to, its achievements, and now in late middle or old age, they are eager to set down their memories before those fade and are lost to recall forever.

Out of the Cradle

The space age had its roots in the twilight years of World War II, with the development of the A-4 ballistic missile, later named the V-2. One of these was launched to an altitude of 100 kilometers above Earth (and hence reached outer space) at Peenemünde on the Baltic on June 20, 1944. Developed under Wernher von Braun, who as a young man became enamored with the idea of interplanetary travel as a member of the Verein für Raumschiffahrt (Society for Space) and launched small experimental rockets from a field in Berlin, these small ballistic missiles had by the end of the war evolved into the brutal Vergeltungswaffen 2 (V-2)—the Retribution Weapon 2—that terrorized cities in Britain, Belgium, the Netherlands, and France.

Following Germany's surrender in May 1945 and Japan's in August, the empires that had divided the world among themselves at the beginning of the twentieth century were broken or destroyed. The United States and Soviet Union remained as the only "superpowers" still standing. At the moment, the United States, relatively untouched behind its oceans and not having had to fight a war on its own territory, had the upper hand. Von Braun surrendered to the U.S. Army to evade capture by the Soviets and brought some 120 of his engineers with him. Helmut Gröttrup and Fritz Karl Preikschat went over to the Soviet side. In the end, the U.S. advantage proved decisive: as Soviet cosmonaut Alexei Leonov told American astronaut Bill Anders in explaining why the Americans had won the race to the Moon, "Your Nazis were better than our Nazis."[2]

As the war ended, both superpowers found that they were at least a decade behind the Germans in rocketry. Joseph Stalin ordered the release of all their own rocket experts from Soviet prisons, including Sergei P. Korolev, who had been rounded up in one of the notorious purges and now took charge of the Soviet rocket program. The United States, having captured the cream of Nazi rocket scientists, had also seized enough V-2 hardware

2. William A. Anders to William Sheehan, pers. comm., July 2018.

to build some eighty of the missiles, which it began testing, assembling, and launching for high-altitude research at White Sands, New Mexico. Though both von Braun and Korolev knew that missile development for military purposes was the highest priority of their respective governments, personally they were far more interested in using them for space travel.

Their dream was, of course, as old as Jules Verne, H. G. Wells, Konstantin Tsiolkovsky, and Robert Goddard—not to mention Lucian of Samosata. Suddenly, space travel no longer seemed so farfetched. Indeed, for many youngsters growing up in that era, flights to the Moon and planets seemed just around the corner. The imminence of our destiny in space was underscored by realistic movies, some of which featured astonishing art by Chesley Bonestell and others. Bonestell's first foray into space art had been a series of paintings of Saturn as seen from each of its then-known moons, published in *Life* magazine in 1944, including *Saturn as Seen from Titan* (figure I.1), "the painting that launched a thousand careers."[3] These and other paintings, mostly made for science fiction magazines, appeared in the book *The Conquest of Space* (New York, Viking, 1949).[4] In addition, Bonestell created huge backdrop paintings of the lunar landscape for the 1950 movie *Destination Moon*. The movie seems rather tame by today's standards but was much admired by famed science and science fiction writers Isaac Asimov and Arthur C. Clarke.[5] It also proved formative for several contributors to the present book.

Von Braun was among the greatest influences on the generation of future space scientists. In addition to his engineering prowess, he was a public relations and marketing genius (and successful, with the help of NASA and the Defense Department, in hiding the extent of his Nazi involvement during the war, which only emerged after his death in 1977). Von Braun, Willy Ley, and several others wrote "Man Will Conquer Space Soon!," a series of articles lavishly illustrated by Bonestell, Fred Freeman, and Rolf Kelp for the magazine *Collier's* between 1952 and 1954, which introduced the concepts of winged ferry rockets, anticipating the later space shuttles; of a spinning wheel-shaped artificial-gravity space station in Earth orbit; of space telescopes tended to by astronauts; and of bases on the Moon (figure I.2). At the time, however, humanity's ultimate adventure seemed to be a voyage to

---

3. Ron Miller and Frederick C. Durant III, *The Art of Chesley Bonestell* (London: Collins and Brown, 2001), 47.

4. Willy Ley, *The Conquest of Space*, illus. Chesley Bonestell (New York, Viking, 1949).

5. Isaac Asimov, *In Memory Yet Green* (New York: Avon, 1979), 601.

FIGURE I.1 Chesley Bonestell, *Saturn as Seen from Titan*, 1944.

Mars, the planet on which the existence of some forms of life was considered probable.

The means of how to go about it had already been outlined by von Braun in a book called the *Mars Project*, written to relieve his boredom while he was interned at White Sands. Its ambition was breathtaking: it called for a flotilla of spaceships to be assembled near the Earth-orbiting space station, each able to carry seventy humans via a 260-day minimum-energy Mars transfer orbit to the red planet. So comprehensive were von Braun's plans that he even pondered the possible psychological problems that might be encountered on the long trip to Mars: "At the end of a few months, someone is likely to go berserk. . . . [If] somebody does crack, you can't call off the expedition and return to Earth. You'll have to take him with you." He proposed censoring radio communication to prevent the crew from hearing dispiriting news from Earth.[6] Needless to say, many adolescents at the time were simply

---

6. Quoted in David S. F. Portree, *Humans to Mars: Fifty Years of Mission Planning, 1950–2000*, SP-2001–4521 (Washington, D.C.: NASA, 2001), 3.

FIGURE I.2 Chesley Bonestell, *Assembling the Ships for the Mars Expedition*, 1956.

stunned by the concept of visiting Mars. So vividly was the human destiny in space imagined that there seemed no doubt as to its fulfillment in the rather near future. Von Braun himself predicted that the first human Mars mission would take place in 1965. That such a date could be taken seriously shows that—despite the Cold War and other existential threats to the planet— futuristic scenarios were not then as uniformly dystopian as they have since become. In the wake of World War II and demonstration of the staggering destructive power of atomic weapons in Japan, it seemed that humans had learned their lesson, and that organizations like the United Nations and an international system of banking would help mitigate international conflicts. The mantra was "Never again," and humans proposed to swear off all that conflict and—with confidence in the genius of the species and belief in the inevitability of progress—were eager to rush forward to a world of possibilities in the larger universe beyond the Earth and to finally venture beyond Tsiolkovsky's cradle of mankind.

Rivals

Meanwhile, in the real world of rocket development, the early fifties saw the U.S. Air Force, Army, and Navy all competing to launch an artificial satellite into Earth orbit during the upcoming International Geophysical Year (IGY), 1957–58. By demonstrating the superiority of its rockets, the United States meant to curb decisively Russia's military ambitions. Technologically, the Soviet Union was regarded as (and was) rather backward. Alarmed at the lack of strategic parity with the United States, the Soviets in 1954 began secretively developing an intercontinental ballistic missile (ICBM) that would be capable of delivering a three-megaton nuclear charge able to reach the United States.

Though believing itself to have no real competition, the United States was to some extent in disarray, with each of the three branches of U.S. military working on its own proposal. Though the air force later developed what became the Atlas rocket, it did not yet have anything suitable. Von Braun, with the U.S. Army, wanted to launch the first satellite on a modified Redstone ballistic missile (a direct descendant of the V-2), which became known as the Jupiter C. In retrospect this was the strongest contestant. The navy's system, based on the civilian Viking and Aerobee rockets, not ballistic missiles, was deemed less likely to alarm the Soviets and was given the green light to proceed in July 1955 as Project Vanguard. Launch was targeted for September 1957 during the IGY (July 1, 1957–December 31, 1958), but because of many delays, it fell behind schedule. This created an opening for the Soviets, and they seized it.

The Soviets had successfully tested the world's first intercontinental ballistic missile on August 21, 1957. This was the R-7, a two-stage rocket with a maximum payload of 5.4 tons (enough to carry their hefty hydrogen bomb from the Baikonur launch facility to the mainland United States). As something of an afterthought, they capped this achievement with another, when Korolev arranged to have Sputnik 1 placed on top of another R-7 and successfully launched it into orbit on October 4, 1957. Though carrying a simple payload, an 83.6-kilogram polished metal sphere with external radio antennas able to broadcast radio pulses (bleep-bleeps) easily picked up by ham radio operators, Sputnik shocked the Western world, which immediately scrambled to catch up.

Clearly, a new era had begun—one that could happen only once, in which humanity's image of the solar system would develop from a crude sketch to the detailed picture that is still evolving to this day.

Amateur Astronomers Still Have a Role to Play

In direct response to Sputnik, the National Defense Education Act was passed by Congress and signed into law by President Dwight D. Eisenhower in September 1958. Its intent was to "ensure trained manpower of sufficient quality and quantity to meet the nation's national defense needs" and to help train a new generation of math and science teachers, scientists, and engineers. In tandem with this was a growing demand for information about space travel and the various bodies of the solar system.

It is now almost impossible to imagine just how exciting it was for those of us bitten by the astronomy bug at the time, when so much of the solar system was still unknown compared to now and when the promise of adventure lay ahead rather than behind us. A key difference was in how information was disseminated. Prompt access to vast amounts of data via the internet—like the latest images from rovers on Mars or orbiters around Jupiter—would have seemed pure science fiction then. After all, television was still a novelty, and for many with a budding interest in astronomy, the main source of information remained the astronomy shelf at the local library, often through books that were years, even decades, out of date. It's no exaggeration to say that two developments at the time that galvanized amateur astronomers and nurtured the careers of many who went on to become professional planetary scientists were the publication of *Sky & Telescope* magazine and the emergence of prolific astronomy author Patrick Moore.

Created through the merger of two money-losing journals, *The Sky* and *The Telescope*, the new magazine, *Sky & Telescope*, began publication under legendary founders Charles A. and Helen S. Federer at Harvard College Observatory in November 1941 (a month before Pearl Harbor). Even with the interruption of the war, the magazine was immediately successful. The goal was to create as large a community as possible of amateur astronomers in the United States and to put them in touch with one another. As editor-in-chief, Charles was a perfectionistic, hands-on, exacting taskmaster, who insisted that his staff be well informed about astronomy and that they get things right, which they consistently did. The magazine thus set a high standard from the start.

This hands-on approach fit well with the do-it-yourself and hobby movements of previous decades, when many readers took up the challenge of grinding mirrors and building their own telescopes. In addition to encouragement for those who wished to do-it-themselves, *Sky & Telescope* also published advertisements for complete telescopes, including Edmund Scientific, Criterion Manufacturing Company, A. Jaegers, and Cave Optical. Such ads launched the dreams of a thousand would-be professional astronomers, including some of those included in this book (figure I.3).

Patrick Moore, a self-taught English amateur who had served as a navigator with the Royal Air Force during the war, came on the scene with tremendous enthusiasm in the early fifties as a leading member of the British Astronomical Association (BAA), the premier amateur organization devoted to systematic studies of the Moon and planets. A practical astronomer, though without academic training in astronomy or geology, Moore specialized in studies of the Moon well before the first spacecraft reached it, producing book after book, including *Guide to the Moon* (1953), *Guide to the Planets* (1955), *Earth Satellites: The New Satellite Projects Explained* (1955), *The Planet Venus* (1956), *Guide to Mars* (1956), and *The Amateur Astronomer* (1957). In 1957 he received a contract with the BBC for a run of six astronomy TV programs to be called *The Sky at Night*. (Television was still a great novelty, and though TV sets were very expensive, seven million homes in Britain already had one.) The timing of Moore's first broadcast was impeccable. When the program first aired in April 1957, a lovely naked-eye comet, Arend-Roland, was gracing the northern sky, and in the autumn of that same year, he discussed the implications of the launch of Sputnik. He later recalled the excitement of those early days:

> Remember, in those days the Space Age was very young—it started only in 1957, and there were still many people who classed interplanetary exploration with science fiction. This meant that the pure observers were very much to the fore. . . . We were, for example, energetically mapping the Moon and, if anyone wanted a detailed view of, say, the crater Clavius, he was apt to turn to a Lunar Section publication rather than to anything else—if only because even the best photographic lu-

FIGURE I.3 An 8-inch Cave Astrola reflector being readied for a night's work by an eager young amateur astronomer, Patrizia Brasch. Courtesy of Klaus Brasch.

nar maps being produced in America still showed less detail than could be revealed by a good telescope, and were very defective in the libration areas. It seems strange to look back at this attitude now, when we have the close-range views from the various manned and unmanned space vehicles.[7]

Moore always emphasized that there was still useful work to be done by anyone with enthusiasm and a reasonably good telescope. This might involve mapping limb areas on the Moon (an effort that was perfected by Bill Hartmann only a few years later) or looking for transient lunar phenomena, monitoring the ashen light of Venus or timing central meridian transits of cloud features across the disk of Jupiter to monitor the planet's zonal winds. Such projects were promoted in his books and by BAA observing sections, as well as by the Association of Lunar and Planetary Observers (ALPO), founded in 1947 by Walter Haas, then a number cruncher at White Sands missile range and a mathematics teacher at the University of New Mexico in Albuquerque.

These organizations helped inspire many amateur astronomers to acquire observational skills, and at least a few made important discoveries—such as the four-day rotation of the upper atmosphere of Venus (by Charles Boyer, by profession a magistrate in the French Congo) and the hurricane-like internal rotation of the Great Red Spot of Jupiter (discovered by Elmer J. Reese with only a 6-inch reflector).

Professionals Join in the Quest

Professionals increasingly began to dominate solar system studies, though amateurs still contributed. A preeminent role was played by Dutch-born astronomer Gerard P. Kuiper, who in 1955, while at the University of Chicago's Yerkes Observatory, became extremely interested in the Moon and more generally in the rigorous application of astrophysical methods to the bodies of the solar system. Kuiper was already a respected professional and, among other important discoveries, had detected methane in the atmosphere of Titan in 1944. Nevertheless, he respected amateur work, and wrote in 1961:

---

7. Patrick Moore, "The Nineteen-Fifties: The Space Age," in *The British Astronomical Association: The Second Fifty Years*, 15–16, BAA Memoirs, vol. 42, part 2 (London: BAA, 1990).

The phenomenal growth of astrophysics and the exciting explorations of the Galaxy and the observable universe led to an almost complete abandonment of planetary studies. . . . Physical observations of planetary surfaces, particularly Mars, led to controversies and speculations that may have been appreciated by the public but hardly by the professionals. More and more this branch of planetary work, including the study of the Moon, became the topic *par excellence* of amateurs—who did remarkably well with it. The *Memoirs* of the British Astronomical Association became the chief record for the development of planetary surface markings. Astronomers with large telescopes were so occupied with the engaging problem of stars, nebulae, clusters, the Galaxy, and the universe that astronomy became almost entirely the science of the stars.[8]

Kuiper's interest in the Moon had been largely stimulated by Ralph Baldwin's book *The Face of the Moon*, published in 1949, which argued convincingly for a meteoritic impact theory of the origin of the Moon's craters.[9] Yet, not everyone agreed with Baldwin's analysis—Patrick Moore, for one, continued to hold forth for a volcanic theory of the craters' origin—and it was clear that a great deal was yet to be done. In retrospect, a crucial date was August 1955, when the International Astronomical Union (IAU) held its annual meeting in Dublin, Kuiper was determined to champion lunar research in general and his own views about the Moon in particular. There was no entirely satisfactory map of the Moon at the time. A Welsh amateur and civil servant had just finished work on the largest hand-drawn Moon map (300 inches) ever made, but unfortunately, H. P. Wilkins was a rather maladroit draftsman, and on the reduced scale at which the map was reproduced (in Wilkins and Moore's *The Moon*, published in 1955), it is so cluttered with fine detail (much of it fictitious) as to be almost unintelligible. Kuiper realized the need for a good photographic lunar atlas; at the IAU meeting he sought to generate interest, and well aware that the greatest expertise in such matters at the time lay with members of the BAA Lunar Section, he succeeded in enticing Alan Lenham, Ewen A. Whitaker, and (later) D. W. G.

8. G. P. Kuiper and Barbara M. Middlehurst, eds., *Planets and Satellites* (Chicago: University of Chicago Press, 1961), vi.

9. Ralph B. Baldwin, *The Face of the Moon* (Chicago: University of Chicago Press, 1949).

Arthur to come to Yerkes to work with him on this and other projects. Whitaker (after clearing the usual bureaucratic and professional hurdles) arrived at Chicago's O'Hare Airport "just as the headlines of the evening newspapers were blaring, 'Sputnik I orbits the Earth.'"[10]

Whitaker and Arthur were assigned to select fourteen-by-seventeen-inch prints of the best lunar photographs available (most were taken at the Mc-Donald Observatory in Texas, but others were from Yerkes and Pic du Midi Observatory in France). Soon running out of allocated space for this, Kuiper converted the east attic at Yerkes (known as the "battleship" because of its porthole-like windows) into a suite of sizable offices, though when even this did not provide enough room, numerous portable tables were set up in the cavernous basement of the dome. Amid all this, Kuiper—now serving his second stint as observatory director—was fighting for his professional life. His tenure was "brief and stormy." According to Yerkes historian Donald E. Osterbrock:

> Kuiper was a great scientist who found it very difficult to distinguish between his own interests and those of his department and of astronomy. He sincerely believed that the research he was doing was the most important there was. Why else would he be doing it? It followed that all the resources of the university should be devoted to it. Why waste time on less important projects? He was always surprised if his colleagues did not see eye to eye with him. . . . Nevertheless, he had become an outstanding planetary astrophysicist, who had reawakened the whole field of solar-system research and made it once more a growing, vital scientific area.[11]

By the end of 1959, it was becoming clear that "all was not well at Yerkes. Undercurrents of discontent were circulating among some of the nonlunar-oriented personnel. The trouble undoubtedly arose from the generally strong-arm tactics that Kuiper used in promoting and favoring the Lunar Project over the more traditional fields of stellar, galactic and extra-galactic astronomy."[12]

---

10. Ewen A. Whitaker to William Sheehan, pers. comm., July 2015.

11. Donald E. Osterbrock, *Yerkes Observatory 1892–1950: The Birth, Near Death, and Resurrection of a Scientific Research Institution* (Chicago: University of Chicago Press, 1997), 312.

12. Ewen A. Whitaker, *The University of Arizona's Lunar and Planetary Laboratory: Its Founding and Early Years* (Tucson: University of Arizona, 1986), 8.

The solution was Kuiper's departure to the University of Arizona, which recruited him to set up, with the Institute of Atmospheric Physics, what would become the Lunar and Planetary Laboratory (LPL) in Tucson. In addition to Whitaker and Arthur (Lenham having meanwhile returned to England), Kuiper gathered around him a group of talented graduate students, including Dale Cruikshank and Bill Hartmann. Chuck Wood joined as an undergraduate research assistant two years later.

The individuals discussed thus far entered solar system studies mainly through astronomy. Others with different backgrounds, however, such as geology and chemistry, were also attracted to the field. The first application of stratigraphy to understanding geological units on the lunar surface, for instance, was the work of geologist Eugene Merle Shoemaker, who in 1959 published what was immediately hailed as the definitive study of the formation near Winslow, Arizona, earlier known as Coon Butte, which had been famously misidentified by noted U.S. Geological Survey (USGS) geologist Grove Karl Gilbert as a steam maar, implying volcanic origin. Shoemaker, fresh from studying craters formed by underground nuclear explosions at the Nevada test site, realized that the crater (now known as Meteor Crater) must have been formed from an enormous blast like those produced by nuclear bombs and thus had to have an extraterrestrial origin. Though doubters remained (they always do), Shoemaker effectively proved Baldwin's thesis that lunar craters had been formed by similar impacts, and in a classic 1960 study of Copernicus, the ninety-seven-kilometer-wide "monarch of the Moon," demonstrated that no volcanic process could ever have produced an explosion capable of throwing out millions of tons of material across two hundred kilometers and more, and that the loops and chains of satellitic craters circling Copernicus must be secondary impacts, caused by debris thrown out at much higher energies than the thick and rugged ejecta that had cascaded over the crater's outer flank.

Shoemaker was instrumental in forming the astrogeology branch of the USGS at Menlo Park, California (later moved to Flagstaff, where it remains today). In 1962, he published the first geological map of the region around the crater Kepler, in which he attempted to analyze the precise stratigraphic sequence of lunar formations. He, Baldwin, Kuiper's student Hartmann, and others posited the likely occurrence of impact craters on other bodies, as was subsequently borne out in spacecraft images of Mars, Mercury, the Galilean satellites, and others. Impacts were also shown to be key in the accretion of the planets from the solar nebula some 4.6 billion years ago. As Hartmann and his collaborator Donald R. Davis initially suggested in 1974, 30–50 million years after the solar system formed, a Mars-sized body apparently collided with the primordial Earth to produce the debris from which the unusual Earth-Moon system had its birth.

This was, of course, the era of Apollo, in which virtually limitless budgets were allocated for space-related initiatives. Despite the scale of expenditures, the scientific exploration of the Moon was almost an afterthought. Again, Shoemaker was the crucial figure, and he spent much of 1963 in Washington, D.C., lobbying for such a science agenda. He faced fierce opposition from the Office of Manned Space Flight, which was only interested in beating the Russians, but in the end he prevailed. Former USGS geologist Don E. Wilhelms has written, "If Shoemaker had not gone to NASA Headquarters to lobby for geology, . . . it is entirely possible that we would have no samples or photographs from the lunar surface."[13]

By now, there was a great deal going on, and other groups formed with a somewhat different take on lunar and planetary astronomy. The California Institute of Technology (Caltech), best known for its stellar and extragalactic studies with the giant telescopes at Mount Wilson and Palomar Observatories, now began to play a leading role in large part because of the presence of the Jet Propulsion Laboratory (JPL), which in 1958, with funding by NASA, became a federal laboratory managed by Caltech. In 1960, it began undertaking research in many facets of planetary science. Another important group was founded in 1965 when geophysicist Frank Press left Caltech for the Massachusetts Institute of Technology (MIT). He later renamed the Geology Department the Department of Earth and Planetary Sciences, hiring several scientists with expertise in the rapidly expanding field. One of these was Tom McCord, who set up the MIT Planetary Astronomy Laboratory (MITPAL) and established a highly open and productive environment, where associates ranging from undergraduates and graduate students, like Clark Chapman, to postdocs and professors (notably the cosmochemist John Lewis) were able to creatively follow new initiatives. A significant number of professional planetary researchers got their start in MITPAL before McCord moved the group to another center for planetary astronomical and geological research, Hawai'i, which offered both excellent telescopes and examples of active volcanism.

---

13. Don E. Wilhelms, *To a Rocky Moon* (Tucson: University of Arizona Press, 1993), 58.

Because of MIT's proximity to Harvard and Brown University, collaborative research often involved those institutions. The Manned Spacecraft Center (now the Lyndon B. Johnson Space Center) made Houston the scientific center for the Apollo program, and it engaged in a long period of collaborative research with Houston's Lunar Science Institute, or LSI (now known as the Lunar and Planetary Institute). Jim Head, who had obtained his PhD at Brown in 1969, provided geological training to the Apollo astronauts, then spent a brief stint as interim head of LSI before returning to Brown to establish a robust planetary geology program there. Also during this period (late sixties to early seventies), Gene Simmons, after serving as chief scientist of the Manned Spacecraft Center, returned to MIT with the entire collection of Moon photographs obtained by the five unmanned lunar orbiter missions between 1966 and 1967. And so, very rapidly, academic centers of planetary science began to flourish beyond the early centers in Flagstaff, Tucson, and Pasadena.

Those were heady days. Though defining the beginning and end of any historical period is somewhat arbitrary, it is not unreasonable to suggest that this golden age lasted from the late fifties (with Sputnik and the Luna missions to the Moon) to the late eighties (when Voyager 2's flyby of Neptune in 1989 concluded the first spacecraft survey of the major planets). For those who participated in it, including some of the contributors to the present volume (and hundreds more), there was simply too much going on to fully appreciate the extraordinariness of the time. Despite the Cold War, Vietnam, resistance to civil rights, loss of faith in religion and in government leaders, there was still room for intellectual playfulness—the exhibition of the "healthy animal spirits of the mind," which American historian Richard Hofstadter suggests is the pursuit of truth, which, like the pursuit of happiness, is better than the possession. He writes, "Truth captured loses its glamor; truths long known and widely believed have a way of turning false with time; easy truths are a bore, and many of them become half-truths. Whatever the intellectual is too certain of, if he is healthily playful, he begins to find unsatisfactory. The meaning of his intellectual life lies not in the possession of truth but in the quest for new uncertainties. . . . The intellectual is one who turns answers into questions."[14]

---

14. Richard Hofstadter, *Anti-intellectualism in American Life: The Paranoid Style in American Politics* (New York: Library of America, 2020), 35–36.

This book attempts to gather a few glimpses of an era whose partici-
pants are inevitably dying out and which will soon be lost to living memory.
Each contributor to this work has had their own motivations and followed
their own trajectory in getting where they did, but for all their differences,
they all shared, in greater or lesser degree, what French writer Antoine de
Saint-Exupéry in *The Little Prince* described as a "yearning for the vast and
endless sea."

En Route to the Planets

WILLIAM K. HARTMANN

I somehow materialized on planet Earth in 1939 in the modest town of New Kensington, Pennsylvania, about twenty miles up the Allegheny River from Pittsburgh. My father was a civil engineer at the Aluminum Company of America, working mostly on uses of aluminum alloys at the company's lab a few blocks from our house. His father was a painter of landscapes and houses; he had emigrated from Switzerland in the 1890s. My mother had taught math before her marriage. Her father had started out as an Alabama preacher and ended up as the head of the Math Department at the University of Illinois, interested in philosophy. As far as I could tell, she and both my grandmothers were happily engaged (as per the paradigm of their day) in managing home life, supporting their husband's careers, and raising their kids. The bottom line is that I was lucky (in my view) to be raised with a mix of practicality, interest in nature, creativity, and philosophizing.

During visits to Carnegie Museum in Pittsburgh, I saw ancient Egyptian and Greek sculptures, and during visits to my maternal grandparents' farm in Illinois, our family found beautifully flaked, centuries-old arrowheads. As a result of such experiences, I developed an interest, at around age eight to eleven, in archaeology and ancient history. At the same time I began to read imagination-expanding science fiction, which in those days was a literature of ideas about possible future societies and exploration of the universe (as opposed to current emphasis on mere fantasy). One day I found a map of the Moon in one of my older brother's books, and I was captivated by the thought that the Moon was not just a light in the sky, but a real *place*, with named mountains and plains. My interest began to shift to our solar system environment, and one day, around age eleven or twelve, I bought a copy of *The Conquest of Space*, by expatriate German rocket expert Willy Ley. The book is still widely known for its amazing paintings of various planets and spaceflight, by the father of American space art, Chesley Bonestell. It's striking how many engineers who put Apollo on the Moon had that book when they were kids. Bonestell later illustrated the 1952–54 series of *Collier's* magazine articles by expatriate German rocket expert Wernher von Braun, explaining how we could explore space. As a result of such literature, all of us fourteen-year-olds, circa 1954, "knew" that we humans would soon be exploring the solar system (although the older generation viewed this as childhood fantasy). Bonestell became a boyhood hero of mine, and I began trying to create my own drawings and paintings of other planets. Miraculously, I became friends with Bonestell two decades later.

Sometime around 1951 I found in a Pittsburgh bookstore, on a remainder table, Ralph Baldwin's now-famous 1949 book on lunar craters, *The Face of the Moon*. It was a bit beyond my abilities at the time but is now credited with proving that lunar craters were caused by meteoritic impact explosions, not volcanic processes. It was a great help later.

Amateur and Early Telescopic Observations

In the summer of 1954, as I turned fifteen, I acquired a copy of *Guide to the Moon*, by the English astronomy popularizer Patrick Moore. It offered a history of telescopic observations of the Moon and recounted various mysteries about lunar surface features. (The grandest telescopes of that time could resolve lunar features only down to about a mile in diameter.) I wanted to start seeing those mysterious features myself. I soon had a 2.4-inch commercial refractor telescope, and my dad helped me construct a 3-inch diameter reflector telescope (with a commercially sold mirror.)

I soon subscribed to *Sky & Telescope* magazine, joined a local astronomy club (Allegheny Valley Amateur Astronomers), and began to learn more about telescope making. I joined the Association of Lunar and Planetary Observers (ALPO), an amateur group patterned on the British Astronomical Association, or BAA. It was run by Walter Haas, a math professor at the University of New Mexico. In those days, professional astronomers focused on stars and galaxies and had little use for the Moon and planets; a joke among professional astronomers was that the Moon was a nuisance, periodically illuminating the sky and interfering with observations of faint galactic features. Thus, it was the amateurs in the ALPO and the BAA, a global network of backyard observers, who, in the journals of those two societies, provided the best documentary records of changes in the mysterious dark markings of Mars, new storm clouds developing on Jupiter, faint dusky features on Venus and Mercury, and so on. In the pages of the ALPO journal, I encountered the names and observational drawings of fellow teenagers, such as Dale Cruikshank and Clark Chapman, both of whom, within a few years, would become admired professional pals. (See their chapters in this book.)

During my last years in high school, with the tutelage of other amateur astronomers, I ground an 8-inch diameter mirror in our basement workshop and built an 8-inch telescope (with Dad's help on the aluminum tube). During those years I sent many drawings of lunar craters and the changing features of Mars, Jupiter, and so on to the ALPO, and I was much encouraged when some of them were published in the ALPO journal, along with similar drawings from other observers.

Moonwatch, Allegheny Observatory in Pittsburgh, and College at Penn State

One summer day in 1955 (age sixteen) I picked up the morning *Pittsburgh Post-Gazette* on our porch and discovered that President Dwight D. Eisenhower, on July 28, had announced that the United States, as part of the International Geophysical Year (which ran eighteen months, from July 1957 to December 1958), would attempt to launch an artificial satellite into orbit around Earth. The project would be called Vanguard, and it would involve a new, *peaceful* rocket program, separate from America's military intercontinental ballistic missiles (ICBMs).

Finally! There it was, the beginning of what all my friends and I, at around age fourteen, knew would come, the era of space exploration. Our parents had ignored such nonsense, but we had been following Bonestell's paintings, von Braun's articles, and our beloved science fiction.

Launching satellites presented an interesting problem: no high-tech satellite tracking systems existed in those days. If a satellite was released from a launch vehicle somewhere above the atmosphere, its detailed orbital characteristics (if it succeeded in attaining orbit) were not immediately known.

To solve the problem, the famous comet astronomer Fred L. Whipple, at the Smithsonian Astrophysical Observatory (Cambridge, Massachusetts), proposed in 1955–56 that amateur astronomers could help locate the satellites and establish their orbits. Operation Moonwatch would establish teams from amateur astronomy clubs in cities worldwide. Teams would gather at Moonwatch stations and sit on a carefully determined north-south meridian line on either side of a pole, with small telescopes distributed by the Moonwatch program. The telescopes were aimed at the top of the pole, and everything was arranged so that the line of observers would see overlapping fields of stars along the meridian. Satellites were typically launched into west-to-east orbits. Operation Moonwatch officials would announce a time interval (maybe forty-five minutes) when a new satellite *should* pass over your Moonwatch station (*if* it had reached orbit). Given a cloud-free sky, one or two of your team would be sure to record the position and time of satellite crossing the meridian.

Our neighboring amateur astronomy club, in Pittsburgh, established such a Moonwatch station on the roof of Pittsburgh's venerable Allegheny Observatory. In 1956–57, I participated in setting up the station atop the observatory, as documented in figure 1.1.

★　★　★

I graduated from high school in spring 1957 (still no satellites in orbit). Where to go to college? I had learned that if you wanted to go into astronomy, it was best to concentrate first, during the four years of college, on basic physics, and then learn serious astronomy in graduate school. So I enrolled in physics at Penn State University, where my brother had gone a few years before. It is located in State College, Pennsylvania, in a central mountain valley that was, as people said, "equally inaccessible from all parts of the state."

But before I went off to college, my dad suggested that maybe, as a high-school graduate, I should be thinking about a summer job. Hmmm. Radically new idea. With his support, I drafted letters to Buhl Planetarium and Allegheny Observatory, both in Pittsburgh, asking if I could have a summer job.

I soon received a letter from Nicolas Wagman, the director of Allegheny Observatory, proposing an interview. Wagman was a wonderful, empathetic man, and he agreed to hire me as a summer assistant. I commuted twenty miles, as far as Pittsburgh, with some of Dad's friends who worked there, and

Bill Hartmann sights through home-made telescope

Ken Hi Student On Team
To Track Earth Satellite

By JEANNE MECKEY
News Feature Writer
"Operation Moon Watch"
satellite tracking teams

"There's 12 men on a team and
a captain," Bill continued. "We'll
by be stationed on the roof of Alle-
may gheny Observatory in Pitts-

scope for his Leica camera.
Both his telescopes are kept in
his father's (E. C. Hartmann)
garage, in the back yard. Any

FIGURE 1.1 Portion of an article in the New Kensington, Pennsylvania, *Valley Daily News*, March 8, 1957, showing the author at seventeen with the front end of his 8-inch reflector telescope, describing his part in the Pittsburgh Moonwatch team. His mom's handwriting is at the top. Courtesy of William K. Hartmann.

then the rest of the way by bus to Riverview Park, where the observatory is located. At this time Allegheny Observatory specialized in a fifty-year-old program of photographing stars thought to be nearby and measuring their *parallax*. Parallax is the tiny angle (a fraction of a second of arc) by which a nearby star's position appears to shift relative to background stars when viewed six months apart from opposite sides of Earth's orbit. Easy analogy: hold your finger (nearby star) in front your face; view from first one eye, then the other. The finger appears to move relative to the background (far wall of room or distant scene—the background of distant stars). The angle of shift is the parallax. The greater the parallactic shift, the nearer the star.

Dr. Wagman put me to work on a routine program of "reducing" measurements from photos that they took, on glass plates. The position of the target star was measured with microscopic accuracy on the plate, relative to the position of four or five distant background stars. This was, mind you, in 1957–61. No computers. The final analysis of the positions was made on a "calculator"—a fifties-era machine on which I punched in each measured star position on a keyboard and added it to the previous measurement by pulling a crank handle protruding from the right side of the machine. The star positions were observed and reduced over several years until enough measurements were available to calculate the target star's distance. In addition to that work, I led tours through the observatory and showed visitors celestial sights, such as craters on the Moon and the rings of Saturn, through the observatory's 13-inch refracting telescope. Eventually I was allowed to stay overnight at the observatory and use the 13-inch to my heart's content, and I also learned from Wagman how to take parallax photos with the 30-inch Thaw refractor (named after a donor who supported its installation).

When I reached Penn State University, in the fall of 1957, I joined their local Moonwatch team as soon as I arrived. One Saturday morning I came down from my room to the adjacent dining hall to discover the *New York Times* (sold in the dining hall) with a stunning headline: the Soviet Union had launched the first artificial satellite of planet Earth, Sputnik 1, on the previous day, October 4! The fact that the communist Russians beat America into space, seemingly out of nowhere, caused huge angst throughout the United States, with the result that federal budgets for science education and science research, especially at NASA, shot way up. This would, of course, prove to be fortunate for me and my fellow science students. It was not so much that I had cleverly arrived at the right time and place, but rather that the right time and place had, by chance, come to me. Russia followed up with Sputnik 2 in November.

I was soon spending evenings and predawn mornings at the State College astronomy club Moonwatch station, watching for the Russian satellites. (The middle of the night was no good for watching early satellites, because the Sun would be on the far side of Earth from the satellite, so no sunlight would illuminate the object in orbit.) State College, Pennsylvania, was not the balmiest place to watch satellites during winter college semesters. Usually the passage of the satellite could be predicted within perhaps twenty minutes, but we had to be set up on time, and then put everything away, so on a typical evening or morning, we might end up sitting or moving around in the dark, in snow, in twenty-degree Fahrenheit weather for an hour or more.

Amid the national mortification following the two Sputnik launches, the first attempt to launch an American satellite, with the U.S. Navy's Vanguard rocket, occurred on December 6, 1957. In contrast to Russian secrecy, ours was an "open" society (at least about some things), and we allowed full national television coverage. The rocket rose a few feet, lost power, fell back vertically onto the launch pad, and exploded in a ball of fire. Von Braun's army team was then given the go-ahead to make an attempt and successfully launched America's first satellite, Explorer 1, on January 31, 1958. Later, the Vanguard rocket would eventually partly redeem itself with three successful satellite launches of its own (out of eleven attempts).

*   *   *

I graduated from Penn State in 1961 with a degree in physics, and the question that year was where to apply for graduate school. I wanted to study the Moon and planets, but the problem was that the field of planetary science had not yet been invented. Astronomer Gerard Kuiper, however, known for his 1944 discovery of the atmosphere of Saturn's large satellite, Titan, and for studies of Mars and the origin of the solar system, had moved from the University of Chicago to the University of Arizona, in Tucson, where he was establishing the new Lunar and Planetary Laboratory (LPL). So I applied to the Astronomy Department at U of A (and a few backup places, which I now forget). My letter of application emphasized my telescopic observations, my work at Allegheny Observatory, and my Moonwatch experience, as well as the idea that NASA was already moving American science toward studies of the solar system. The timing (no thanks to me) could not have been better. On April 12, charismatic Russian cosmonaut Yuri Gagarin became the first human to orbit the Earth in space. Worried about the loss of American prestige, on May 25, President John F. Kennedy proposed to Congress that the United States should "commit itself to achieving the goal, before this decade is out, of landing a man on the Moon and returning him safely to Earth."

Graduate School, Gerard Kuiper, and Discovering the Orientale Impact Basin on the Moon

At the University of Arizona, I cobbled together a scholastic program in the Astronomy and Geology Departments that might produce an education about planets. In the meantime, Kuiper, across the street from the Astronomy Department, in what we called the Lunar Lab, put me to work on an interesting project of his, to create a "rectified atlas" of photos of lunar formations as seen from overhead (which might help with the Apollo program, when astronauts would be viewing various parts of the Moon from overhead). With the assistance of English lunar expert Ewen Whitaker and English cartographer D. W. G. (Dai) Arthur and others, Kuiper had assembled first-generation copies of the best lunar photos from the best observatories. Then, in a separate campus building, they had set up a system where these photos were projected down a long hall onto a three-foot-wide white hemisphere representing "our half of the Moon," that is, the Moon's near side, visible from time immemorial (the far side had been revealed for the first time in fuzzy photos by the Soviet Luna 3 spacecraft in October 1959). Kuiper's rectified atlas would divide the near side of the Moon into sections, and my job was to orient the camera and the projected photos so I could photograph each section "from overhead" with a "4 by 5 camera" (using four-by-five-inch high-resolution film), as shown in figure 1.2. Then I would take those films to the Lunar Lab darkroom (across the hall from Kuiper's office) and print them on large photographic paper so they could be published in the atlas.

The Moon wobbles slightly during its orbits around Earth so that actually a little more than half (59 percent) is visible at one time or other, and sometimes one edge or another can be seen especially well. One day, I walked into the room where a photo with a particularly good view of the Moon's east edge was projected onto the globe. In my recollection, our technical assistant, Harold Spradley, who handled the glass plates going into the projector, and perhaps one or two other students were standing around, discussing two or three strangely curved concentric ranges of mountains visible along that eastern edge. They saw them as a geological curiosity, but having read Baldwin's book about lunar impact features (back home as a teenager), I realized that the mountains formed a large concentric ring structure, and I immediately recognized them as the visible portion of what was likely to be the youngest and best-preserved lunar impact basin visible from Earth. Some

FIGURE 1.2 The author placing four-by-five-inch film in the camera to photograph an image of the Moon's front side projected onto a hemisphere, 1961 or 1962. Courtesy of William K. Hartmann.

lunar mountains, such as the Rook Mountains, had been mapped in the area, and a relatively small patch of lava, known for several centuries, had been named Mare Orientale by German observer Julius Franz in 1906. Because these features were at the edge of the Moon, however, they could be seen from Earth only at a very shallow angle, and the early mappers had thus not realized that the mountains formed a huge and beautiful set of concentric rings around the un-prepossessing lava patch Mare Orientale. Actually, the most interesting Orientale lava, to me, was not the patch in the center but smaller patches lying along the bases of the cliffs constituting one of the mountain rings. (They indicated that the cliff ring formed with deep fractures [geologists call them *faults*], which allowed molten lava from the Moon's interior to squeeze up to the surface and erupt along the base cliff ring.)

I quickly departed from my assigned task of photographing predefined sections of the Moon and took photos of this formation. I then shared prints of the photos with Kuiper, who agreed that it was an important discovery and graciously let me be first author on our 1962 paper, announcing the discovery of the well-preserved Orientale impact basin. In preparation for the paper, I made a series of overhead photos of other major impact basins, demonstrating that most other, older impact basins (more beat up by subsequent impacts) had traces of the same concentric ring pattern—a pattern of cliff-forming fractures that must relate to the lunar crust's response to the largest impacts. Our paper was published in *Communications of the Lunar and Planetary Laboratory*, a series of publications started by Kuiper, which ran from 1962 until 1973, the year of Kuiper's death. The idea that our Orientale basin images showed part of a larger, well-preserved concentric ring feature was confirmed five years later, in 1967. Then in 1971, Charles A. Wood, who was an undergraduate at this time (see his chapter in this book), published an article surveying thirty-one impact basin structures on the two sides of the Moon, as well as the transitions between various smaller "ordinary" crater structures and the huge multiring structures.

I felt proud to be publishing my first "professional" paper with the famous Gerard Kuiper! More importantly, I learned a significant if subliminal lesson about science. Namely, that sometimes it pays to back off from the subject at hand. In those days, everyone else in lunar science seemed to be concentrating on the smallest possible details that could be seen on the Moon (especially in the sixties, in preparation for Apollo astronaut landings). But we had backed off from the Moon (literally as well as figuratively) and managed to discover a thousand-kilometer-wide feature that had not been recognized before. Focusing on finer and finer detail is not the only way to make discoveries in science.

And there was another, more subliminal lesson. Why hadn't others recognized these concentric ring patterns? Baldwin (unaware of Orientale) had already approached that idea, for example, emphasizing the mountain rings extending in short arcs partway around Mare Imbrium, but his work was still regarded as controversial. I began to realize from books I was reading at that time, such as *The Mentality of Apes*, a 1921 book by Wolfgang Köhler, that the answer had to do with gestalt psychology, which relates to what we call pattern recognition today. Gestalt is a German word referring to the brain's ability (or inability) to recognize a large-scale pattern as a whole, as opposed to being distracted by individual parts. This recognition often comes by a sudden insight rather than a piece-by-piece analysis of details. Köhler's book described chimpanzees faced with a banana hanging out of reach, in a room full of boxes, moving around frustratedly, but then *suddenly* recognizing the solution of piling the boxes to get the banana. In other words, they experienced sudden insight about the overall situation, rather than focusing on details. In the same way, the sixties obsession with the smallest possible lunar details may have inhibited most lunar scientists from seeing how things fit together into a significant larger pattern.

I encountered the gestalt problem more dramatically a month or so after Kuiper and I published our findings, when I got a letter from Nobel prize–winning geochemist Harold Urey. I was thrilled, as a mere grad student, to hear from such a famous scientist—until I opened the envelope. Urey indicated that he could not see the concentric symmetry we had found in the various lunar impact basins. (He had earlier written about axial symmetry in the large Imbrium basin). He pointed out to me something more useful and serious: that Kuiper's series of LPL *Communications* was not peer reviewed. Summarizing, he advised that this work, of which I was so proud, was not a good way to start a career—not a message my lowly graduate student self was happy to receive from a Nobel Prize winner.

In any case, the concept of multi-concentric-ring structure among the largest features was quickly and widely accepted by the lunar geology community, much to my pleasure. And I now think that Urey's complaint about publishing without peer review (other than Kuiper's approval) was valid. In Kuiper's defense, however, the *Communications* was a means by which he and other LPL scientists could circulate their own results (often consisting of data compilations too long to publish in a journal). No doubt there was an underlying bit of personal psychology in this story as well. Urey and Kuiper had famously feuded about many issues. Perhaps this feud was a factor in Urey's letter. Whatever its motivations, the letter was a generous communication with an unknown student.

The problem of recognizing patterns in nature began to fascinate me. It relates to an issue familiar to field geologists. There is an old proverb, "If I hadn't seen it, I would never have believed it." That seems an obvious statement, and a key to the scientific method—the idea that an event or phenomenon must be "seen" or recorded in some repeatable way to be accepted as reality. Amusingly, however, some witty field geologists often say, "If I hadn't believed it, I never would have seen it." This seemingly paradoxical statement relates to the key fact that field geologists often establish the existence of some major feature, like a massive but eroded fault crack in Earth's crust, through detailed mapping of rock formations at various local sites. Only then, by backing off to some mountaintop vantage point, *after the feature is known to exist*, is the fault crossing the landscape easily seen, when it might have been missed before. This is truly a case of "If I hadn't known about it, I never would have seen it."

All this led me to issues that have fascinated me ever since. Are some scientists (and people in general?) genetically programmed to be more sensitive to essential, logical details, and others to more "backed off," broader, perhaps less rational, perceptions? This relates, of course, to the research of left brain and right brain activity, with evidence suggesting that the left brain is more involved in language and logical progressions, and the right brain with visualizing things in terms of broader pictures—although current research suggests that this idea has been overplayed. For what it is worth, Carl Sagan, around 1986, sent me one of his books with an inscription saying that he thought I had a good corpus collosum (a nerve bundle connecting the left and right hemispheres)!

Helping to Create Mauna Kea Observatory in Hawai'i

A new adventure materialized in 1964. Twenty years earlier, Kuiper had already discovered that Saturn's satellite Titan, the largest moon in the solar system, had a thick atmosphere. (Moons had typically been dismissed as airless bodies.) Kuiper had observed the spectrum of colors in sunlight reflected from Titan and discovered that certain colors (wavelengths) had been absorbed (blocked) by methane gas—a Titan atmosphere. He had always wanted to extend these observations, but water vapor in Earth's atmosphere blocked certain crucial colors that would reveal other gases. As a result, Kuiper had always been interested in creating state-of-the-art observatories on Earth's highest accessible peaks, which would be above most of Earth's water vapor.

To solve this problem, Kuiper had focused on the summit of Mauna Kea volcano, a 14,000-foot-high mountain on the volcanically active, biggest island in the Hawaiian chain. It was the highest peak that was easily accessible by car, and it rose above most of Earth water vapor (despite being in the middle of the Pacific Ocean). Kuiper employed Alika Herring, who had previously worked as a professional telescope maker, and together, in spring 1961, they installed one of Herring's 12-inch-diameter reflectors in a dome on the summit of Mauna Kea (figure 1.3). Kuiper sent Alika to Hawai'i to use the telescope to get records of the seeing conditions atop Mauna Kea—for example, to check very close together double stars, to see if the atmosphere was stable enough for the telescope to resolve them into two separate objects.

In June 1961 it was time to give Alika a rest at home, so Kuiper assigned me to replace him for six weeks. Here I was, having grown up in the deciduous woodlands of Pennsylvania, and having just moved to the cactus-strewn Sonoran Desert of Arizona, now sent off to the alluring palms and beaches—or rather, wait—I mean, the 14,000-foot peak of Hawai'i. I recall the airlines trying to capitalize on the historically friendly "aloha spirit" of Hawai'i and hiring attractive young women to meet airplanes at the Honolulu airport, drape passengers one at a time with a beautiful flower lei necklace, and bestow (at least on the males) a nice kiss on the cheek. For some reason, this custom has disappeared. Having said that, I must praise the true aloha spirit of the islands in the sixties. For example, merchants at small stores would always direct you to another store if they didn't have an item that you sought. The whole atmosphere of Hawai'i was friendly and helpful.

Alika met me and showed me around the interesting town of Hilo, at the foot of Mauna Kea, before driving me up to the isolated stone ranger cabin, Hale (pronounced "hall-lay") Pohaku. The Hawaiian term means "house of stone." I would live here alone, at about 9,000 feet, not counting my one-hour drives every few days into Hilo for supplies (see figure 1.3). (As far as I know at this writing, Hale Pohaku still stands, about one hundred yards behind the present-day Mauna Kea Observatory visitor center.) Each clear night I would drive to the top of Mauna Kea on a crude dirt road (where even contemporary astronomers have had occasional substantial accidents driving off an edge).

Kuiper and Alika had worked out an arrangement with a ham radio operator in Hilo, so that I would call in from the walkie-talkie in the jeep as I left Hale Pohaku, when I reached the top, and when I was back safe at the cabin. Theoretically, someone would show up to rescue me if our Hilo contact did not hear from me. I'm glad my mother didn't know the details of this crazy arrangement.

With Kuiper's usual gracious paternal interest, he urged me, as part of my education, to spend my occasional spare time learning about the amazing array of natural features of Big Island (the local name for the biggest, and only volcanically active, island in the Hawaiian chain). Most important to

FIGURE 1.3 *Left*, Alika Herring and our trusty jeep at the dome housing Herring's telescope atop Mauna Kea. *Right*, Herring introducing the author to Hale Pohaku, the stone cabin at 9,000 feet on Mauna Kea. Courtesy of William K. Hartmann.

me was Hawai'i Volcanoes National Park. Back in Arizona my grad student pals and I had driven several hours south to visit the Pinacate volcanic region in Sonora, Mexico, where we had seen our first examples of lava flows and nearly mile-scale volcanic craters. But here in Hawai'i was similar *active* volcanism, with smoke fumes rising from craters only a few years old. Another interesting place was the meteorological observatory on the slope of Mauna Loa, the second highest volcano on Big Island, where (being in the middle of the Pacific Ocean, away from local pollution sources) Charles Keeling had, since 1958, created the best historical record of the twentieth-century rise in our planet's atmospheric CO_2. Of course, Keeling's discovery of rising CO_2 was crucial to our understanding of global warming, but the program was not well known at the time. Since Hawai'i had to import all its fossil fuels, I was also able to see fledgling attempts to harness geothermal and wind energy. In short, Hawai'i was full of efforts that the rest of the world is only now coming to appreciate.

I fell in love with the Hawaiian Islands, where I would return often in later years, teach classes in planetary science, and establish a relationship with the Volcano Art Center, in Hawai'i Volcanoes National Park. The center specialized in art related to the national park and sold several of my paintings of that area (figure 1.4). Decades later, I was able to introduce my family and grandchildren to Hawai'i, one of Earth's many interesting landscapes and cultures.

Crater Chronometry

Back in Tucson, during my graduate assistantship in the early sixties, another task was to help construct the LPL's catalog of lunar craters, measuring the sizes of all lunar craters down to four kilometers in diameter on most of the near side of the Moon. As I helped in measuring crater diameters in the early sixties, I began to recognize some interesting aspects about the population of craters on the Moon. First of all, if the craters were created by asteroid impacts, as was now becoming clear, then the number of craters per square kilometer in different regions would give quantitative measurements of the *relative* age of each surface. The more craters per square kilometer, the older the surface. Second, if we knew something about the *rate* at which craters were formed, we could get a first quantitative estimate of the *absolute* age of lunar surface features. Third, the catalog provided a chance to measure the number of impacts at each crater diameter, in other words, to measure

FIGURE 1.4 Painting by the author showing the view from the west rim of Kilauea crater (foreground), in Hawai'i Volcanoes National Park, with the 2008 smoke plume from the smaller pit crater, Halema'uma'u, on the distant floor of the caldera. Courtesy of William K. Hartmann.

the *size distribution* of craters. A key idea here was that while the smallest craters might not survive as long as the biggest, deepest craters, which would give at least the relative age of formation of a surface (such as a lava flow), the smallest might reveal the time scale of erosional processes, such as the destruction of small lunar craters by later, shallower flows in local areas.

Still a graduate student in 1964–65, I applied these ideas to estimate the age of the lunar lava plains, where the low crater density made it easy to create a complete list of craters developed since the lava surface had formed. (In other words, in the lava plains, we did not have to worry about the problems of crater overlap and saturation that applied in the heavily cratered lunar uplands.) But how could anyone estimate the rate of crater formation to allow calculation of the "absolute" ages of the lunar lava plains? In the oldest region of North America, the so-called Canadian Shield, Canadian scientists in the late fifties and early sixties were discovering and dating numerous ancient impact craters. I assumed that they had pretty good counts and radiometric dates for the largest craters, so I used that information to estimate the crater production rate in the Canadian Shield. Since Earth and Moon had been companions in the same part of the solar system almost since the system formed some 4.5 billion years ago, I assumed that both had had about the same impact rate. As a result, I published a 1965 paper in the planetary science journal *Icarus*, estimating that the average age of the lunar lava plains was about 3.6 billion years. This publication was four years before the Apollo astronauts returned the first lunar rock samples, which dated various lava plains in the general range of 3.8 to 3.2 billion years. This good match gave some predictive validity to the scheme of using crater counts to measure planetary ages.

I have to add an amusing story. My friend and colleague Don Wilhelms, a widely respected lunar scientist and raconteur, wrote a book, *To a Rocky Moon*, describing the history of lunar studies. In that book he described my correct prediction of lunar lava plain ages, but he added, "I have always thought that Bill has led a charmed life, although he may be just plain smart." I've replied that I was never sure which option was true.

A related epiphany occurred one day in 1964 as I was driving down Mauna Kea. The first successful American probe to reach the Moon, Ranger 7, designed to take photos as it deliberately crash-landed on July 31, returned images of many craters, as small as about ten meters in diameter, much smaller than had ever been observed from Earth. The epiphany was that I could use these images to extend the size distribution of lunar craters down to much smaller sizes than had been measured before, allowing a relatively complete size distribution for lunar craters (and hence giving information about the size distribution of objects hitting the Moon). This might reveal erosion rates of the smallest craters (for example, by the sandblasting effect of micrometeorite impacts.) This work would thus reveal human-scale properties of the lunar surface, as visited by astronauts.

*   *   *

After getting my PhD in 1966, I began writing popular-level articles about planetary science, in journals such as *Smithsonian, Natural History, Sky & Telescope,* and *Astronomy.* Some of the articles included my attempts, inspired by Bonestell's paintings (not to mention my Swiss grandfather's Earth-landscape paintings hanging around the house), to paint my own views of what distant worlds might look like, based on the newest discoveries by the slowly growing number of planetary scientists.

In the seventies my wife, Gayle, had relatives in San Francisco, and this led to an opportunity to meet my boyhood hero, Chesley Bonestell, who lived with his wife, Hulda, just down the California coast in Carmel. Chesley was a wonderful old fellow, in his eighties, who took us zipping up the steps to his backyard studio. He pretended to have a gruff exterior and loved to find flaws in the paintings of other artists. For example, Chesley chortled that one version of Norman Rockwell's painting for the U.S. postage stamp celebrating the first Apollo lunar landing showed sunlight illuminating the lunar landscape—but Earth, up in the lunar sky, was illuminated from a different direction! (Aristarchus of Samos, around 250 BC, had realized that the Moon was illuminated by the Sun and used the geometry of lunar phases to prove that the Sun was much farther away, and much larger, than the Moon—but this was forgotten a few centuries later, when Ptolemy's Earth-centered solar system was adopted until Copernicus came along. It's interesting that most people even today don't know of Aristarchus's simple argument.) Chesley's critiques about the direction and color of lighting in planetary paintings were helpful to me in my attempts at astronomical art. Chesley and Hulda remarked, on one of our visits, that they called me "Hurricane Bill," because "I always blew in with a Gayle."

I found that the process of making astronomical paintings required synthesizing everything known about the given world (whereas scientific work often required dissecting knowledge according to individual techniques, such as spectroscopy, geology, or theoretical models of orbital history). The fun was to try to figure out what a scene in the solar system would look like if you could be standing there (figure 1.5). In ways difficult to describe, the painting process often suggested to me new tractable issues that had not been fully studied, such as sky colors, or textures of surfaces at a human scale, or the nature of the surface of a comet nucleus. When you try to paint a planetary surface, you quickly realize which things you know about that planet and which things you don't know.

Meanwhile, I continued the work of developing a system using crater populations to estimate the history of planetary surface features. I called the technique *crater chronometry*. It was a great time to begin working on planetary science, because only handful of researchers were working in that area. One day around 1970, the phone rang in my assistant professor office at

FIGURE 1.5 *Looking Homeward*, a 1981 painting by the author, showing the lunar surface at sunset. The distant crater rim (and the Earth) is illuminated by sunlight, but the foreground is illuminated only by bluish light from fully lit Earth. Courtesy of William K. Hartmann.

the University of Arizona (in the Lunar Lab, now called the Kuiper Building). On the other end was Bruce Murray, involved with the Mariner space probes to Mars (and later head of the JPL). He was wondering if I would like to be on the imaging team of the upcoming Mariner 8 and 9 spacecraft projects, which would attempt to send the first spacecraft to enter orbit around Mars and map the red planet from there (figure 1.6). After a millisecond of consideration, I agreed.

Mariner 8 failed soon after launch, but Mariner 9 headed off successfully—chased by the Russian Mars 2 and 3 combined orbiter and lander spacecraft. Mariner 9 was the first to enter orbit in November 1971. Mars 2 and 3 followed soon after (and Mars 3 released a lander that, upon reaching the surface, stopped transmitting after only twenty seconds). Mariner 9 was the only real success among the three missions, making the first complete orbital map of Mars. What an exciting experience! During our extended mission in 1971–72, Mariner 9 made many spectacular discoveries, revealing a new

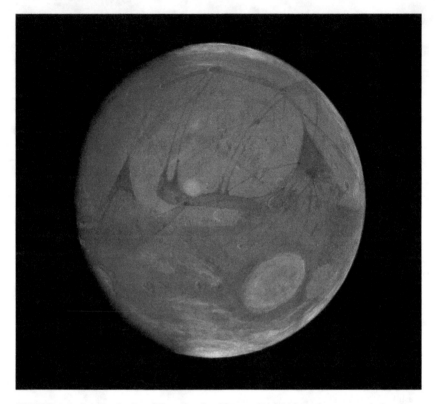

FIGURE 1.6 Painting by the author showing Mars as it might have been seen during the approach of Mariner 9 if Percival Lowell had been correct. The view shows the system of straight-line "canals" as shown in Lowell's maps and telescopic drawings. In reality, only irregular, changeable, streaky dark markings exist in some of the "canal" positions, probably related to wind-blown dust. Courtesy of the William Sheehan collection.

Mars: the huge shield volcanoes, dry riverbeds, and dune fields so familiar today. My role on the team was to apply crater chronometry to estimate the ages of surface features on Mars. The crater production rate on the Moon had now been measured, using surface ages based on lunar rock and soil sample dating supplied by Apollo astronauts (as well as samples returned by Russian robotic missions beginning with Luna 16 in September 1970). For Mars, we estimated a higher cratering rate than for the Moon, because it was closer to the asteroid belt, source of the asteroid impactors. Using those data, my Mariner 9 papers, starting in 1973, suggested that the youngest Martian lava flows (the most sparsely cratered ones) could be as young as a few hundred million years, much younger than those on the Moon. This result was controversial, because flyby missions prior to Mariner 9—Mariners 4, 6, and 7—had, by chance, obtained several snapshots of many heavily cratered parts of Mars, which had led to the view that Mars was like our Moon, geologically inactive, with little if any recent volcanism, but with some wind to blow the dust around.

By the early eighties, however, a certain class of meteorites falling on Earth were identified as rocks that had been blasted off Mars during asteroid impacts. The decisive clue as to their Martian identity was that they contained pockets of gas exactly matching the now-measured composition of the Martian atmosphere. These meteorites included igneous rocks dating from 1.4 billion years ago as well as other rocks apparently only a few hundreds of million years old. The crater chronometry system had once again been verified, meaning that dating samples from various regions of the Moon and Mars could continually improve the system so that surface ages and the geological evolution of planets, Moons, and asteroids could eventually be clarified without expensive missions to sample each individual geological formation.

Later Work

Meanwhile, starting in the late seventies and continuing through the eighties and nineties, I was invited to write textbooks on planetary science and general astronomy, which went into multiple editions. I also published popular books illustrated with my own paintings and those by my friends Ron Miller and Pamela Lee, as well as (so far) two novels, trying to deal in a more freewheeling way with human exploration and relationships. I began to feel more and more that it was important to share the larger picture of our science with the public. After all, we are dealing with the relationship of humanity to the universe. These sorts of writing, and the paintings, led to my 1998 award of the first Carl Sagan Medal for Excellence in Public Communication in Planetary Science from the Division of Planetary Sciences of the American Astronomical Society.

★　★　★

During those same decades of the eighties and nineties, my late German colleague Gerhard Neukum developed a more organized program of crater chronometry, which involved what we might call "German precision," sometimes using stereo pairs of images to count the craters. I suspect that he achieved greater precision than I did in terms of crater density. When it comes to ages, however, the measuring transfer of measured cratering rates on the Moon to estimated cratering rates on Mars probably involves an uncertainty of around a factor of two. That may sound terrible at first—until we realize we are dealing with orders of magnitude in ages. Crater-count dating can certainly distinguish 100 million-year-old Martian surfaces from 3 billion-year-old surfaces, and that difference is crucial when interpreting the geological history of Mars. Gerhard and I were sometimes at odds in terms of our techniques, but I'm happy to say that during a 1999–2001 team project at the International Space Science Institute (ISSI) in Switzerland, the late, jovial ISSI director Johannes Geiss insisted that Gerhard and I jointly write a chapter in the ISSI book *Evolution and Chronology of Mars*. We did, and despite our differences, we agreed on the basic issues. (Mostly because of that joint authorship, Gerhard and I shared the 2002 Runcorn-Florensky Medal from the European Geophysics Union, which I treasure as an international award.)

Also in the eighties, I reunited with my graduate student pal Dale Cruikshank (see his chapter in this book) in projects that brought me back to Mauna Kea, in Hawai'i, where my lonely cabin at Hale Pohaku was superseded by a beautiful dormitory, and where the summit now bristled with multiple large telescopes from various countries. Mauna Kea was now one of the leading observatories in the world. Dale had taken a position at the University of Hawai'i, which meant that he was allowed observing time on some of the Mauna Kea telescopes. Dale and I were both interested in the surfaces of bodies in the outer solar system. His expertise was in telescopic instrumentation and spectroscopic analysis of such bodies, as well as in measuring their *light curves* (changes in brightness, caused by the rotation of elongated bodies or, less often, by bodies having different brightness on different sides). My role was to think of problems that could be clarified by such observations. We had several successes. One involved Halley's comet. We had discovered that among outer solar system bodies, reflectivities were correlated with their visible and infrared "colors." We drew up an empirical diagram showing reflectivity versus color of outer solar system bodies where these two properties had been measured. Our diagram showed that bodies already known to have very bright ice surfaces (like Jupiter's satellite Europa) were slightly bluer than ordinary sunlight, while bodies that were already known known to be dark tended toward a redder chocolate color. As the astronomical world prepared for comet Halley's trip through the inner solar system and around the Sun in 1985–86, our friend, the late astronomer Mike Belton, bravely attempted to summarize everything we humans thought we knew about Halley's comet. At the time, comets in general were believed to be, in Fred Whipple's words, "dirty snowballs." Belton cited a fairly bright reflectivity of 28 percent as the favored value for Halley. Our measurements at Mauna Kea Observatory, however, soon after Halley was recovered, combined with our diagram, indicated a much lower reflectivity of only 4 percent, which we published in 1985. Dale tells the story of presenting our result at a comet conference in Flagstaff, Arizona, and being roundly criticized because "everybody knew" that colors of objects had nothing to do with reflectivities. But then, the European Giotto probe made a close pass by Halley's comet nucleus and measured a reflectivity of 4 percent. Dale reports that a journalist who had been at the Flagstaff meeting approached Dale about a year later to ask, "How does it feel to be proven right?"

Observing comets from Mauna Kea Observatory reinvigorated my curiosity about what it would look like to be on the surface of a comet nucleus. I attempted some paintings (e.g., figure 1.7). Exhibiting such paintings at scientific meetings, circa 1992, I asked for critiques from comet scientists. I was interested to find that most comet scientists during that period had given virtually no thought to the geology or appearance of the surface on a comet nucleus, because their observational techniques were limited to spectroscopy of dust and ions in the comet tail, which formed a halo of dust and gas that hid the nucleus.

★　★　★

In the last few years I've been even more interested in the big picture of how our scientific work is done and transmitted to the public, and about how people relate to one another (humans being perhaps the biggest mystery of all). This is all the more important since we, in the 2020s, live in a world where polls indicate that about 35–40 percent of Americans (but not most other nationalities) believe that our Earth was created only six thousand years ago and that biological evolution did not occur. But that's another story, which I'm addressing in another (as yet unpublished) book manuscript.

Let me conclude with a thought about careers. It's good to find a subject you are interested in, but it's also good to find a subject that offers prospects

FIGURE 1.7 View from the dark side of a comet, painted by the author. In this view we look down the streamers of the tail of the comet (opposite to the direction of the Sun), so that the shadow of the comet nucleus is cast on the dust tail of the comet. Courtesy of William K. Hartmann.

for a career. I was lucky in being present at a time and place when, despite excellent work by amateurs, strikingly few astronomers were studying the planets, and moreover, the solar system was beginning to be explored for the first time by spacecraft. As a result, my graduate student friends and I could help create the emerging field of planetary science. As the first generation of planetary scientists, we were presented with a solar system full of low-hanging fruit, and it seemed that everywhere we looked, discoveries were just waiting to be made. Inevitably, it's harder for young planetary scientists today, with limited budgets for research despite large numbers of young researchers. Nonetheless, endless opportunities exist for exciting, fruitful work, for amateurs and professionals alike. Amateurs are still doing some of the best work in monitoring changes in the clouds and surface features of Mars as well as the cloud formations on Jupiter and Saturn. A related field involves the crucial understanding of the stability of our own planet's environment and ecosystem, since civilization itself is threatened by climate change and water shortages. Additional challenges are searching for past or present biological activity on Mars, probing for liquid oceans under the icepack surfaces of some satellites, determining the nature of planetary systems being discovered near other stars, and most tantalizing, searching for evidence of life (even intelligent life) on Earth-like planets near other stars. And let us not forget the need for well-informed science journalists to keep the public correctly informed about our understanding of the universe, in an age of increasing sources of misinformation on social media.

Pathway to Pluto

DALE P. CRUIKSHANK

I was born in Des Moines, Iowa, in 1939, just three weeks before the German invasion of Poland that marked the start of World War II. In my earliest years, I lived on Des Moines's east side with my grandfather and two great-aunts. Webster Elementary School was about two and a half blocks from the house, and I walked to and from school every day, often in the snow. I was helped across the busy street in front of the house, and from there on to school, I was on my own.

It was during my third-grade year that, along the way home from school one day, I discovered a big brick building called the Iowa State Archives. My path down the alleyway behind the building allowed me to peer through windows into the basement, where some chemists in white lab coats were busy with their boiling flasks, microscopes, and other fascinating equipment. I would often stop, crouch down, and peer through a window at the activity below. One day, one of the chemists who had frequently seen me at the window beckoned me to come around to the front door and visit the lab. I eagerly accepted the invitation.

One of the chemists showed me around, explaining what they did for the State of Iowa. They monitored the octane content of gasoline, butter fat concentration in milk, weeds in seeds to be sown, and other things for which there were standards established by the state. I was invited in several more times, and when I received a chemistry set for Christmas that year, I took it to the lab to show my new friends. They tried to explain things like atomic structure, molecules, and valence, but it was a little beyond me at the time.

On my ninth birthday, in the summer of 1948, the chemists invited me into the lab and had a little party for me. As a gift they gave me a set for making little metal soldiers by pouring molten lead into molds. I took it home, melted the lead on a gas stove in the basement of our house where my aunt heated water for the laundry, and made soldiers. I had a lot of freedom to do such things on my own.

Around the age of eleven or twelve, I moved to Urbandale (1950 population 1,777), a suburb of Des Moines, where I lived until I left Iowa after graduating from what is now Iowa State University in 1961.

When I was about twelve, I ran across the book *Sun, Moon, and Stars,* by W. T. Skilling and R. S. Richardson, while browsing in a bookstore for the latest Hardy Boys mystery. I bought that one instead of the Hardy Boys, and it remains today on my shelf as my first astronomy book. Among many other things, I learned about solar eclipses, and I later learned that the path of a total eclipse would cross Minnesota and Michigan, not so far from Iowa, on June 30, 1954. When I was thirteen, I sent away two dollars for a set of plans to make a telescope, with complete instructions for grinding and figuring a 6-inch mirror. My scoutmaster helped me with some of the details, so with a mirror-making kit in hand, but no clear understanding of a parabolic figure, I began grinding the glass. It took forever, but in the meantime, I was making plans to see the 1954 eclipse.

To make the eclipse trip, I had to request a leave from my job at the corner drugstore, where I worked as a soda fountain clerk. I had started work at age thirteen (summer of 1953) for thirty-five cents an hour. By the summer of 1954, I was earning forty-five cents an hour, a princely sum that mostly supported my new photography hobby. To photograph the eclipse from my chosen site in Michigan's Upper Peninsula, I worked with a friend to build a camera around a lens I had acquired. It was the front half of a portrait lens— the Rapid Rectilinear, with a diameter of two inches and focal length of about twenty-eight inches. This gave an almost decently sized image of the Sun on a four-by-five-inch sheet of film.

After a bit of persuasion, my mother agreed to drive my high-school friend Wayne Mansfield and me to the Upper Peninsula, where we arrived on the central path about two days before the eclipse. Wayne and I camped in our tent, prominently emblazoned with "Operation Eclipse, Des Moines, Iowa." The morning of June 30 dawned in a local fog, which we watched slowly thin out toward the time of totality. When that magic time finally arrived, we could see the inner corona fairly well. Our photos weren't so great (I still blame the fog), but I had taken my first lesson in eclipse photography, a subject that I didn't finally conquer until the Great American Eclipse of August 21, 2017.

Operation Eclipse yielded benefits beyond the mediocre photos we took. Charles A. Federer Jr., founder and editor of *Sky & Telescope* magazine, was in our area for the eclipse. With his family in tow, he made the rounds of the amateur astronomers there, and his entourage stopped by to see what Wayne and I were doing. The photo in figure 2.1 was taken by Federer and sent to me a few weeks later. I started my subscription to *Sky & Telescope* as soon as I got home after the eclipse. To date, three *Sky & Telescope* covers feature photos I took.

More adventures occurred in those early years in Urbandale. I worked in the drugstore for five years, played the piano and organ for services in our church, and tinkered with chemistry by making gunpowder and other moderately dangerous things. In the early fifties there was a growing awareness of space and space travel as something more than just fiction. In 1953 and 1954,

FIGURE 2.1 Dale Cruikshank (*left*) and Wayne Mansfield in Michigan at the total solar eclipse of June 30, 1954. Photo by Charles A. Federer Jr. Courtesy of Dale P. Cruikshank.

Collier's magazine ran cover stories about artificial Earth satellites, and the cover of the April 10, 1954, issue showed a painting of Mars with a fleet of spaceships approaching, while posing the questions, "Can we get to Mars?" and "Is there life on Mars?" I certainly wanted to know.

There were favorable oppositions of Mars in July 1954, and especially in September 1956; public viewing on both occasions was held at the Drake Municipal Observatory in Waveland Park, Des Moines. The telescope is a vintage 8.5-inch Brashear refractor made in 1894, with a gravity-driven flyball governor mechanism operating the sidereal drive visible through a window in the pier. At the 1954 apparition, I got acquainted with Philip S. Riggs of Drake University, the astronomer in charge, and who gave the public lectures. Riggs was very friendly to me over the years, and among other things conscripted me to help with public nights for the 1956 Mars apparition. This was my first experience with crowd control—the lines waiting to see the planet through the telescope stretched out the observatory door for a city block or more. When friends and I organized the local astronomy club (later to become the Des Moines Division of the Great Plains Astronomical Society), Riggs kindly made the observatory available for our monthly meetings. In 1957, Riggs wrote a letter of recommendation for me for a summer assistantship at Yerkes Observatory, the start of a chain of life-changing events.

As an aside, in the autumn of 1958, after I had finished my first summer assistantship at Yerkes Observatory, Riggs phoned me one afternoon to say that Seth B. Nicholson, the Mt. Wilson and Palomar astronomer with whom I had corresponded as a kid, was at his home and would like to meet me. Nicholson was a hero of mine, having discovered as many moons of Jupiter as Galileo did, and perhaps more importantly, replying to my letters from time to time. He had even sent me a book about the Sun. Meeting Nicholson that day was a thrill that still brings a smile to me. Perhaps most significant of all, this meeting and other events have caused me to strive to pay forward such kindness to young people who contact me about science, who might be influenced to enter the field someday or are just curious.

The postwar period, with its high-altitude experiments using repurposed V-2 rockets and the buildup to the International Geophysical Year, or IGY (July 1, 1957, to December 31, 1958), has been well documented, while the public appetite for space was whetted with books and magazine articles by Wernher von Braun and others, illustrated with Chesley Bonestell's wonderful paintings of lunar and planetary scenes, not to mention Willy Ley's exceptional *Rockets, Missiles, and Space Travel* (I bought my copy in 1953). As the IGY approached, Carl Gartlein in Canada recruited amateurs all over the continent to participate in an aurora watch, and plans were made for Moonwatch, a network of amateurs with small telescopes set up, to get visual observations of the intended artificial Earth satellites. The Amateur Scientist column in the monthly *Scientific American* described the details of a rocket built by amateurs using readily available materials. I was interested in all these things—aurora watch, Moonwatch, and rockets.

For the aurora watch, I used the official little alidade and paper forms to report what I could see from Urbandale (latitude 42.5° N) from time to time. One night I looked to the northeast and saw a remarkable display of vertical light streaks. I managed to capture the scene on film and promptly sent my best picture to *Sky & Telescope* magazine. They published it as the cover photo on the April 1960 issue and labeled it "unusual aurora." Within days, I received a letter from the editor telling me that Gartlein had seen my photo and pointed out that it was not an aurora at all, but most likely a display of light pillars above bright lights on the ground scattered vertically by ice crystals in the air. Naturally, I felt bad about this misidentification but took some comfort knowing that the *Sky & Telescope* editors and experienced staff, which included at least one professional astronomer, missed the mistake and put the photo on the magazine's cover. This was my first mistake in science, but alas, it would not to be the last of my career. The road to deep understanding in science often has a few blunders and bumps along the way. I have hit my share, but as I sometimes advise young scientists, if you don't make an occasional mistake, you probably aren't working on the cutting edge of your subject.

All of this set the stage for the rocket episode, which formally began with the Amateur Scientist column on rockets in *Scientific American*, which I read religiously. It was probably in 1956 that *Scientific American* published an article about an amateur rocket maker who had compounded his fuel by mixing powdered sulfur and granulated zinc dust in some particular proportion and had successfully flown a rocket a few feet in length. Around this time, I got acquainted with Dick Welch, who was a few years older than me and lived a few doors down the street. He was interested in science, or at least in explosions, and the two of us, with a high-school friend, Bob Curnutt, decided to experiment with rockets, using the zinc-sulfur formula that we had read about. Dick knew where to get zinc dust—from a local paint manufacturer in Des Moines. It was used in a galvanizing paint product, and it was cheap. I got the sulfur through the drugstore where I worked, and with these two prime ingredients in hand, we began experiments to make a rocket.

For the design, we sketched one roughly along the lines of the rocket in *Scientific American.* As a professional draftsman, Dick used our ideas and sketches to produce full-scale drawings of the various parts that we thought we needed. We settled on making the main rocket out of a five-foot length of water pipe about two inches in diameter. The "payload" section of the rocket would be a piece of aluminum tube about three inches in diameter, attached to the top of the main booster. The payload section contained a spring-loaded parachute I already had and a gravity-sensitive trigger mechanism we designed (but that probably would not have worked) to propel the parachute out of the tube at the peak altitude and let the whole thing float gently to the ground.

Without relating all the details of building the rocket, which we named Acamar, I'll skip to the final, or near-final, scene.

To take our rocket project to the next step, we faced two problems— the first was that we had no safe place to launch it, even though Urbandale was surrounded on three sides by farmland. Try as I might, I couldn't find anyone among the many local farmers who would agree to let us set up and test the rocket on their property. We weren't talking about a rocket going to the Moon or an intercontinental ballistic missile (both of which were much in the news in those Cold War years). But even the prospect of a modest five-foot hunk of water pipe rushing through the lower atmosphere was too much for them to get their heads around. In short, we couldn't find a willing host anywhere among the owners of those tens of thousands of acres of cornfields surrounding Urbandale and stretching all the way to the Rocky Mountains. The second problem was that I had graduated from high school in May 1957 and was on track to leave in the fall for Iowa State, so time was running out.

Finally, one day in early September, we decided to grasp the nettle and proceed with a test firing of the rocket in the tomato patch behind my house on 67th Street. It would be just the bare booster stage, strapped securely to a big post. We had a supply of sulfur and zinc, but only enough to fill about one-third of the rocket, so Dick, Bob, and I mixed up what we had, loaded the tube, and strapped it to the post with metal bands.

A touch of the ignition button produced a mighty roar. Instantly, the five-foot length of water pipe, painted white, with "Acamar" in bright red letters, ripped through the two metal straps and streaked into the sky. A column of white smoke stretched at least four hundred feet vertically into the air from our backyard. I can't reliably remember the sound it made—maybe a "whoosh" rather than a "roar," but certainly not an explosion. In any case, the rocket was gone; we couldn't see it attached to the post or even up in the sky, and we had no idea what had happened to it. It was an overcast day, but the clouds were high. As we craned our necks skyward, eventually a tiny dark speck appeared. As it got larger and larger, it became clear that it was our rocket, tumbling slowly, slowly back to Earth, in the general direction of the school building on the other side of the street from our launching pad. We had time to harbor the chilling thought of how much destruction a fifteen-pound length of pipe in free fall might cause to the roof of the new building or to anyone or anything that happened to be inside. Fortunately, the rocket fell harmlessly to the ground, and we managed to recover the hot tube and drag it back home, where the town's vol-unteer fire department was waiting for us, soon to be joined by the town constable (fig-ure 2.2). The story of our unintended rocket launch was widely reported in local and na-tional newspapers. In the following weeks, we were painted by the press as both young ne'er-do-well hooligans and promising youth who would one day take America into the future of space exploration. My father's stern reprimand over this whole event, and the gleam of pride in his eyes while delivering it, helped me take the latter path.

In preparation for tracking the artificial Earth satellites that the United States ex-pected to launch during the IGY, networks of photographic telescopes were set up around the globe, and teams of amateurs were re-cruited to participate in Operation Moon-watch. Moonwatch teams would use simple low-power telescopes arranged along a me-

FIGURE 2.2 *Left to right*, Dale Cruik-shank, Dick Welch, and Bob Curnutt with the rocket Acamar after recovery from its accidental launch in September 1957. Courtesy of Dale P. Cruikshank.

ridian to time the passage of the satellite across that meridian and determine elevation. This information was telephoned in to the Smithsonian Astrophysical Observatory, where it was used in orbit calculations. I tried to organize a Moonwatch team in the Des Moines and Urbandale area but failed to get even the modest financial support needed to purchase the necessary equipment. We tried for a while to make do with empty cardboard tubes instead of telescopes, but it obviously wasn't going to work out very well, and we gave up after a few tries.

As U.S. plans to orbit a satellite moved forward, a couple of limited test firings of the Vanguard rocket had been accomplished. Yet, the United States was famously forestalled when, on October 4, 1957, the Soviet Union successfully launched a rocket carrying the satellite Sputnik 1 into orbit around the Earth. I remember that date very well because it was a Friday afternoon, and I was on my way back to Urbandale from college (Iowa State) about forty miles away in Ames, Iowa, when I heard it on the car radio. The world was shocked, and the effect on the United States and the rest of the world was profound. The world changed that day. The Soviet Union had beaten the United States into space, and the space race that would last decades was on.

The next morning, October 5, the *Des Moines Tribune* headline read, "That Russian Moon—You Can't See It Yet—Maybe It Will Show by October 20."

Later that month, my first-year physics professor, Percy Carr, organized some early morning sessions on the roof of the physics building to try to spot Sputnik with the naked eye. I can't remember seeing it, but there were reports in the newspapers of people on the ground claiming to hear an audible "beep, beep, beep" from Sputnik as it passed over—clearly an example of wishful thinking.

When I started my first year at Iowa State, it was called Iowa State College of Agriculture and Mechanic Arts, but by the start of my second year, it had been renamed Iowa State University of Science and Technology. The university maintained a strong leaning toward agriculture, and even with the new name, it continued to offer courses in tree stump blasting, sheep shearing, and poultry husbandry. None of these courses were central to my career goals, so I concentrated on a curriculum in chemistry, physics, and math. Largely because of the kickstart that the Soviet Union had given to the space age, however, I decided that some proficiency in Russian was desirable. In my sophomore year at Iowa State, I began to take the first-year Russian course taught by a stern but effective language teacher. Thanks to her, a native of Baku who spoke flawless Russian, I grew to be fascinated by the culture as well as the language. Three years of language study with this fine teacher would later serve me well.

During my freshman year, I inquired about the graduate program in astronomy at the University of Chicago, and although I was clearly about four years shy of starting graduate work, I received a reply from Joseph Chamberlain at the Yerkes Observatory. He told me about summer research assistantships at the observatory and encouraged me to apply as an undergraduate. I did so and, in support of my application, provided a letter of recommendation from Professor Riggs, who was still at Drake University. Beyond my wildest expectation, my application was accepted, and so I came to set out for the famous Yerkes Observatory in Williams Bay, Wisconsin, in early June 1958. It was world famous for the 40-inch Clark refractor, which had been (and still is) the largest operational refracting telescope in existence. The mere thought of being in the presence of such an instrument was thrilling beyond words.

I arrived in Williams Bay on a sunny Saturday afternoon, and as I drove my 1953 Studebaker onto the observatory grounds, I saw that a softball game was in progress on the grassy ellipse in front of the building. I later learned that the two teams on the field were the Yerkes Turkeys and the YOYOs (Yerkes Observatory Youth Organization)—faculty members battling the graduate students.

I parked the car and, after a quick look at the decorative terra-cotta stonework on the parapets of the building, went up the steps to the front door and rang the doorbell. Chamberlain let me in, and we went across the foyer to his office for a chat. I was immediately struck by the marble floors, the cavernous hallway leading past offices and the library to the marble steps leading to the 40-inch telescope. Even later, when the offices were occupied, the most prominent sound in the hallway was often the ticking of the two pendulum clocks—one keeping sidereal time, the other solar time. The ticks slowly came into synchrony over the hours and then slowly separated.

In that first conversation Chamberlain told me that I would be working for Joe Tapscott, who managed the darkroom complex in the basement of the building. That was fine with me because I had been interested in photography for years and had taught myself a lot of darkroom techniques at home. Because the observatory doors were always locked, I would be given a personal key, allowing me to come and go as I pleased.

After our conversation, I went back to my car and pondered the moment. Yerkes Observatory makes a profound impression, whether on the first visit or when approaching and entering the building every day for months at a time. The extraordinary architecture, the exterior decoration, the three domes, and of course the 40-inch telescope—all relics of the nineteenth century, and all incongruous with the Wisconsin rural countryside in which the observatory sits.

In the fifties, Yerkes was in its heyday, as astronomer and one-time graduate student Donald E. Osterbrock has described in his history of the institution.[1] The resident staff included S. Chandrasekhar, W. W. Morgan, W. A. Hiltner, J. W. Chamberlain, D. M. Hunten, G. van Biesbroeck, H. A. Abt, K. Prendergast, D. N. Limber, L. Wallace, and others. Many astronomers from Europe and elsewhere came and went, spending a few days or a few weeks in residence. E. J. Öpik visited during that first summer, staying for a few days at the van Biesbroeck boarding house, where I also stayed. Several advanced graduate students were also in residence, enriching both the scientific atmosphere and the social life of the observatory.

When I reported for work the first day, I was given a tour of the building by Elliot Moore, a graduate student in residence at Yerkes. The highlight was first sight of the gargantuan 40-inch telescope. He escorted me up the marble staircase to the massive metal double door, which he opened and then closed with a clank that reverberated through the main hallway. We stepped into the dome with nothing but the telescope pier in sight, standing in the center of what seemed to be an enormous pit. We had entered below the observing floor, and as we climbed the stairs attached to the wall, we eventually came to the floor itself, and above it loomed the telescope. Nothing can compare to the first sight of the 40-inch refractor, a relic of the previous century, built like a battleship, and at sixty-five feet in length, utterly colossal in size. I'm sure Elliot was amused at my reaction. Although at the time I could hardly imagine ever using this telescope, reality later turned out quite differently.

A day or so after I arrived at Yerkes, I met Alan Binder, who also had just finished his freshman year and been given an assistantship for the summer. Al worked in the darkroom complex under Joe Tapscott's supervision, as did

---

1. Donald E. Osterbrock, *Yerkes Observatory, 1892–1950: The Birth, Near Death, and Resurrection* (Chicago: University of Chicago Press, 1997).

I, and we hit it off from the beginning. Al and I had very similar interests, and above all, we were anxious to get a look through the 40-inch. In the meantime, one of the main tasks in the darkroom complex was processing photos of the Moon in preparation for what would become the photographic lunar atlas that Gerard P. Kuiper, the observatory director, was readying under a government contract.

I had first heard of the world-famous astronomer Kuiper in 1954, in a brief newspaper article in the *Des Moines Register* announcing his discovery of a W-shaped cloud on Mars, seen with the McDonald Observatory 82-inch telescope. Here I was now, working in his observatory, and what is more, he also came to the darkroom to help make prints of some of the more important lunar photos. A few times in those early weeks in the darkroom, I found myself standing next to him, hands in the solutions, developing prints and at his instruction rubbing parts of the printing paper with my fingers to bring up the image to the level of density that he wanted. Many years later, after more than a decade of association with him, I was invited to write a biographical memoir of Kuiper's life for the National Academy of Sciences.[2]

Al Binder and I shared a lot of experiences throughout the summer, the most extraordinary of which were the opportunities we had to use the 40-inch telescope. We were both checked out by an experienced staff member, and because the telescope was under little demand, we were given permission to use it alone to look at objects of interest, in both daylight and nighttime hours. On one night, I learned how to measure double stars with a filar micrometer, instructed by George van Biesbroeck, one of the most famous of all double-star observers. On other nights, I watched Elliot Moore take photographs of the Moon and A. P. Lenham expose plates of star fields to measure stellar proper motions.

Al and I had our own interests, principally in the planets, and we set out to observe Venus and Mercury in the daylight hours. Finding Venus was easy, but much fainter Mercury was more of a challenge, even with the telescope's precision-setting circles. At a magnification of 550, and sometimes higher, we could see faint smudges on both objects, which we dutifully sketched, comparing our drawings and notes only after both of us had finished so as not to influence each other.

---

2. D. P. Cruikshank, "Gerard Peter Kuiper, 1905–1973," in *Biographical Memoirs of the National Academy of Sciences*, vol. 62, 259–95 (Washington, D.C.: National Academies Press, 1993).

That summer of 1958 at Yerkes was a life-changing experience. It had been so rich and so much fun that I was sure it was a once-only event. It is hard to describe my astonishment, then, when toward the end of August, as I was about to return to Urbandale, Joe Chamberlain came down to the darkroom to say that I would be welcome to come back the next summer on a similar assistantship. Al Binder was also encouraged to come back for the summer of 1959. We both eagerly accepted this encouragement and returned the next June for another summer of working in the darkrooms and spending spare time at both the 40-inch telescope and the almost equally antique 24-inch telescope carrying one of the first aluminum-on-glass mirrors, which had been made by G. W. Ritchey.

That first summer I was paid six dollars an hour for half-time work, and that was enough to get by with room, board, and incidentals. When the time came to apply for the second summer, I asked if I could work full time, thinking that this would still leave enough time for observing and other activities, while giving me somewhat more spending money. My request was successful, and the second summer I was given full-time employment—for three dollars an hour.

By the time Al and I returned to Yerkes for our second summer of darkroom work, Kuiper had hired two lunar experts, D. W. G. (Dai) Arthur, and Ewen A. Whitaker, both Brits who were well-known members of the British Astronomical Association. Dai Arthur, an expert photogrammetrist, was busy making accurate coordinate grids for the large-scale lunar photographs that Kuiper's project had produced, while Ewen worked in the darkroom, often with help from Al and me, to get the best possible prints from negatives taken at Yerkes, Lick, and McDonald Observatories. All of this would eventually result in three atlases of the Moon, which Kuiper had contracted with the U.S. Air Force to produce in connection with NASA's intention to send people to the Moon.

An additional duty that Al and I shared was to assist with Saturday afternoon visitors to the observatory. The visitors would line up outside the front door and wait for the next tour. When the time came, Al and I would open the door, count the number of people, and admit them to the dome of the 40-inch telescope, where they would hear a brief lecture and see how the telescope moves to point to different parts of the sky. We also manned a little desk where we sold postcards showing photos of celestial objects and of the observatory. One of the lecturers who came to Yerkes to speak to the Saturday visitors was Carl Sagan, at that time still a graduate student.

That Saturday afternoon, while the visitors were all in the dome listening to Sagan, Al and I were at the postcard desk. I was watching and talking as Al was doing a practice calculus problem. Down the quiet hallway came Subrahmanyan Chandrasekhar, walking slowly with his hands clasped behind his back. He saw us at the desk and casually stopped by to see what we were doing. He looked at the problem Al was working on, offered some advice on solving it, and then slowly walked back to his office looking quite satisfied.

I returned to Yerkes Observatory for a third summer, 1960, this time working for Dai Arthur in calculating grid points for what would become the *Orthographic Atlas of the Moon*, published in 1961. Toward the end of that summer, it became known that Kuiper was planning to leave Yerkes and the University of Chicago for another institution. At the end of 1960 and beginning of 1961, Kuiper relocated to the University of Arizona and established the Lunar and Planetary Laboratory (LPL). He offered me another summer assistantship for 1961, and I drove from Urbandale to Tucson in June in my 1953 Ford. Kuiper's entourage in the move to Tucson included van Biesbroeck, Arthur, Whitaker, Moore, editorial collaborator Barbara Middlehurst, and various others who had been part of the Yerkes scene. Details of Kuiper's move and the establishment of LPL are admirably told in Derek Sears's biography of Kuiper.[3]

Al Binder also went to Tucson to join the new LPL, and we both worked in Kuiper's spectroscopy lab, where the infrared spectrometer he brought from Yerkes was being reincarnated for use on the telescopes at nearby Kitt Peak National Observatory and on the McDonald 82-inch in Texas. Al and I learned about infrared spectroscopy and long-path absorption cells that summer. We also met and worked with Tobias (Toby) Owen, a graduate student who had followed Kuiper to Tucson to finish his PhD, and William K. Hartmann, who also came to Tucson to work with Kuiper, and he and I began a lifelong friendship and productive scientific collaboration (figure 2.3). Al stayed on in Tucson, but I returned to Iowa State in the fall for one more term of courses.

---

3. D. W. G. Sears, *Gerard P. Kuiper and the Rise of Modern Planetary Science* (Tucson: University of Arizona Press, 2019).

FIGURE 2.3 *Left to right*, Alan Binder, William Hartmann, and Dale Cruikshank with Dale's 12-inch telescope in Tucson, around 1967. Courtesy of Dale P. Cruikshank.

Returning to Tucson with all my belongings in January 1962, I took a room in a house near the university and began my graduate studies. Al and his wife, Mary, lived nearby, as did Bill Hartmann. Others who arrived on the LPL scene at various times were Charles A. Wood, Floyd Herbert, and Carmelite priest Father Godfrey T. Sill, all of whom became friends, camping partners, and colleagues on scientific papers. Clark Chapman, an undergraduate at Harvard, also came to Tucson several times for summer work at LPL, and together, under the auspices of the Association of Lunar and Planetary Observers (ALPO), we wrote and edited a book-length manual for amateur astronomers interested in the Moon, planets, and comets. As often as I could, I observed the planets with my 12-inch reflector, often in collaboration with Hartmann and Binder (figure 2.4).

In my years as a student at the University of Arizona, I took graduate courses in astronomy, geology, and meteorology, and I assisted Kuiper in the laboratory and at the telescopes at Kitt Peak National Observatory and Mc-

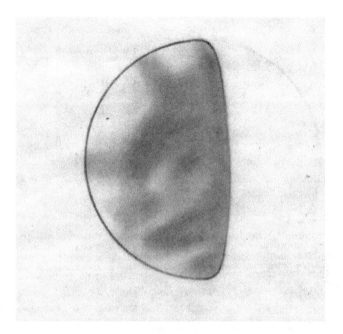

FIGURE 2.4 Mercury, the innermost planet, a composite of sketches by Dale P. Cruikshank based on observations with an 11-cm refractor on January 16, 18, 19, and 20, 1962. At the time, the rotation period was still thought to be equal to that of revolution, eighty-eight days, and markings were identified with those shown on the classic chart by E.-M. Antoniadi from 1934. The names of features, appropriately enough, derived from Greek mythology related to the god Hermes—for example, a large dark area near the south pole was called Solitudo Hermae Trismegisti (Wilderness of Hermes the Thrice Greatest). Some three years after this drawing was made, radio astronomers discovered that Mercury's rotation was 58.65 days, or two-thirds the period of revolution—and data from the visual era had to be drastically reinterpreted. Courtesy of Dale P. Cruikshank.

Donald Observatory. I learned more about infrared spectroscopy and helped build spectrometers that we used in the lab, on telescopes, in a high-altitude balloon, and on a NASA airplane, the Convair-990 (called Galileo), which would crash during a test flight in 1973. Al and I were allowed to use the spectrometers for our own projects as well as in support of Kuiper's observational work, and this resulted in several of our own published papers, as well as some coauthored with Kuiper, all while we were still graduate students. Kuiper also encouraged me to build a field spectrometer for observations of volcanic flames that sometimes occur at Kilauea Volcano in Hawai'i and volcanoes elsewhere. In support of that work on the combustion of volcanic gases, he authorized several trips to Hawaiʻi during a long-term eruption cycle at Kilauea. Ultimately, with David Morrison and Kenneth Lennon, I published a paper (in *Science* in 1973) reporting that the combustion of volcanic hydrogen in air produces blue and yellow flames at some phases of eruptive events.

My association with Kuiper from 1958 to 1970 had a profound influence on my thinking and the way I do science. I have thought back on those experiences many times, and a few additional remarks about my interaction with Kuiper may be of interest.[4]

As a student-assistant in Kuiper's spectroscopy lab, I met with him often and frequently accompanied him to the observatories at Kitt Peak and in Texas. In the course of working with the spectrometers and absorption cells in the lab every day of the week and doing other things of interest to Kuiper, additional projects came to mind. The other students and I, including Al Binder, Bill Hartmann, and later Toby Owen, decided to do some of those things independently. Kuiper readily agreed to let us buy dangerous chemicals and other items needed for a line of observations of experiments relevant to planetary surfaces and atmospheres, including the construction of new apparatuses. He also encouraged our requests to use his equipment on the telescopes at Kitt Peak and on the new telescopes set up at the Steward Observatory Catalina Station on Mount Bigelow in the Santa Catalinas. I am extremely and eternally grateful for the freedom we were given, and I have tried to pass that attitude and available resources on to students over the years.

When wrestling with the numerous difficult issues facing Kuiper as the director of the LPL, he often needed someone to listen to him, a sounding board. It did not seem to matter who that person was, and I frequently found myself in his office listening as he worked through a decision. This usually occurred after hours and on weekends—he would be alone in his

---

4. See also my article "20th-Century Astronomer," *Sky & Telescope* 47 (1974): 159–64, a biographical memoir I wrote at the request of the National Academy of Sciences, as well as a published interview with me by Kuiper biographer Derek Sears, "Oral Histories in Meteoritics and Planetary Sciences—XX: Dale Cruikshank," *Meteoritics and Planetary Science* 48, no. 4 (2013): 700–711, from the supplementary material for which some of these remarks are adapted.

office during those quiet times, and when he wanted to talk about an issue, he would walk over to the lab to see if anyone might be there, or down the hallway to the office that Binder and I shared, to find someone to talk to. Since I was commonly there in the evenings and on weekends, I was often summoned to his office while he talked through a situation, sometimes scientific, but frequently related to personnel and the organization of the lab. Sometimes it would be about his interactions with NASA Headquarters, or with scientists in other institutions. He would typically begin by saying, "This is confidential, but. . . ." I would acknowledge the confidentiality, and he would start talking about things that I often had no business knowing about, or didn't understand, or both.

My PhD research, which I completed in 1968, focused on mineral identification on the Moon using infrared spectroscopy with the LPL 61-inch telescope on Mt. Lemmon. Although the Apollo program to land astronauts on the Moon and bring back samples was well under way, its success was not assured. My identifications of some lunar minerals were basically confirmed but also quickly superseded by the success of Apollo 8 in the summer of 1969. My thesis work was jointly supervised by Kuiper and S. R. Titley, a widely known geologist of the Department of Geosciences at the University of Arizona.

In the last months before receiving my PhD degree (in 1968), I prepared for a ten-month research visit to the USSR under the auspices of the National Academy of Sciences, which had an exchange arrangement with the Soviet Academy of Sciences for scholars and scientists to spend extended periods in one another's country. Kuiper strongly supported me in this endeavor, and in August 1968, en route to Moscow to begin my research work under the National Academy program, I flew to Prague to present a paper on my lunar research at the International Geological Congress convened there. Less than two days after my arrival, the Soviet Red Army invaded Czechoslovakia and took control of the government, all media, and all transportation. I was able to present my paper the next day, but then the International Geological Congress collapsed. Europe appeared on the brink of a major war, and it was unclear if I should proceed to the USSR. After a week in Prague, I was finally able to get a train to Vienna, where I pondered my next step for several more days. In the end, I decided to go on to Moscow; I arrived there on September 1. Two days later, my then wife, a native of Afghanistan, and our two small children (ages eight months and four years) joined me in Moscow. The challenges of setting up a living arrangement in a Moscow residential hotel with two little children, especially in the prevailing tension of the invasion of Czechoslovakia, were great, but we eventually surmounted them.

My host in the USSR was Vasili Ivanovich Moroz, of Moscow State University. He was the leading infrared astronomer in that country, whose work I had followed through publications and correspondence. His work with near-infrared instruments closely paralleled that of Kuiper and, by extension, me. His family and mine quickly became friends. I returned to Moscow in 1973 for five more months of work with Vasili Ivanovich. We had a warm professional and personal relationship until he died in 2004. His daughter, Lyubov Moroz, whom I met when she was five years old, later became a prominent scientist. I have had the pleasure of authoring research papers with her in recent years.

I returned to LPL in the summer of 1969, where I worked as a research associate in Kuiper's lab, while also doing independent research. Then, a research position opened at the Institute for Astronomy (IFA) at the University of Hawai'i, and in the summer of 1970, my family and I packed up and moved to Honolulu. During the seventeen and half years I spent at the IFA working with David Morrison, William Sinton, and Robert Murphy, as well as students Heidi Hammel, Alex Storrs, Robert H. Brown, Nancy Morrison, Terry Martin, and others, Mauna Kea Observatory came of age, with the installation of four major telescopes. At nearly 14,000 feet above sea level, Mauna Kea is a prime site for infrared observations, and it was a great privilege to participate with many colleagues in realizing its potential for stellar, extragalactic, and planetary astronomy. Although I had achieved tenure at the university and the rank equivalent to full professor, in 1987 the opportunity to return to the mainland United States and join NASA as a civil servant scientist drew me to NASA's Ames Research Center, where I continued my work until retirement on July 30, 2021.

I was drawn to Ames by Michael Werner and James Pollack, in part to work on the Space Infrared Telescope, which when finally launched was renamed the Spitzer Space Telescope. Although the project was moved from Ames to the Jet Propulsion Laboratory soon after I arrived in California, it was a great success, and I remained active as an interdisciplinary scientist until its end in 2019. While in Hawai'i, I had become involved with NASA's Voyager project, making small contributions to the scientific discoveries made as Voyagers 1 and 2 flew by Saturn, Uranus, and Neptune (only Voyager 2 flew by Uranus [1985] and Neptune [1989]).

The thirty-two years I spent at Ames were rich in opportunities to learn about astrochemistry from eminent scientists such as Louis J. Allamandola. Lou managed the Ames Astrochemistry Laboratory, which engaged many talented astronomers and chemists over the years. Most of my work on NASA's Cassini mission to Saturn was accomplished while I was employed at Ames, and NASA's New Horizons mission to Pluto and the Kuiper Belt was also conceived and executed during those years. As I write, the New Horizons spacecraft continues its journey deep in the outer solar system, following its spectacular flyby of the Pluto system in 2015 and the extraordinary encounter with Kuiper Belt object Arrokoth in 2019. With this wonderful scientific achievement, and to top off several decades of observations of so many solar system bodies, New Horizons was my ticket to Pluto, the most distant known planet.

At Ames, I also met and married my wife, astrophysicist Yvonne J. Pendleton, who is well known in the United States and Europe for her discoveries of complex organic molecules in interstellar dust in our own galaxy, but also in another galaxy as well. Our scientific interests merge in the outermost solar system, in the thousands of small objects at Pluto and beyond that populate the Kuiper Belt. I study the ices and organics on these bodies, working from inside the solar system outward, and she studies the gas, dust, and ices that arise in interstellar space and accrete to form planetesimals and eventually planet-size bodies. We enjoy working together, and we have jointly authored several research papers reflecting the discoveries we have made, usually in collaborations with other scientists who have mutual interests.

★　★　★

Scientists are in the business of making new discoveries about nature, and my career has offered many examples in which I take pride. Not all discoveries are of equal importance, although they all shed new light on the way the universe works and contribute to the enrichment of the human experience.

For six decades I have pursued the exploration of the solar system, largely from the nearby planets to Pluto and the Kuiper Belt, following developments in sensor technology and the availability of increasingly large telescopes and more capable spacecraft. These developments have enabled constantly improving studies of ever more distant and fainter bodies in the solar

FIGURE 2.5 The author's ticket to Pluto—the Atlas 5 rocket, launched on January 19, 2006, carrying NASA's New Horizons spacecraft. New Horizons arrived at Pluto and flew 12,500 kilometers above the surface of the planet on July 13, 2015. Courtesy of Dale P. Cruikshank.

system, as well as closeup encounters with all the major planets and many comets and asteroids. In a sense, for me the grand prize has been Pluto and its moons, and I have been privileged to see them up close (figure 2.5). But even more distant solar system bodies and everything in between the planets await further exploration by future generations of nimble scientists who are driven to explore and to know more about the part of the universe in which we live.

On the Way to Mars

BAERBEL KOESTERS LUCCHITTA

Many people know from an early age what their interests are, where they want to go in life, what professions are appealing. I was never one of them. Perhaps as a girl I was not expected, nor encouraged, to have ambitious plans. In hindsight, however, I recognize that two principles guided me. One was "go with the flow and grab the opportunities," and the other was "keep an open mind about alternative ideas." Even though I was not aware of it, these concepts served me well. The exploration of space, however, was not in my sight for a long time.

My life started under threatening circumstances. My father, Bernd Koesters, an architect, had settled in a small town in the Rhineland, which was declared a demilitarized zone after World War I. Yet, in 1936, Adolf Hitler, the chancellor of Germany, marched his army into the area. In September 1938, Hitler expanded this aggression by occupying the Sudetenland, a part of then Czechoslovakia, under the pretense of liberating the Germans living there. War was on the horizon. My mother, highly pregnant with me, fled from the dangerous area near the French border to my paternal grandparents' home in the northern German city of Muenster (my father was away on business). The ensuing uproar in the Western world was appeased temporarily with the Munich Agreement on September 30, 1938. On October 2, I was born.

A year later, in 1939, World War II broke out. In 1942 my father moved our family to Weimar, a town of culture and history and the site of the Weimar Republic between the two world wars. Soon thereafter, my father disappeared from my early life: he had been conscripted, and I saw little of him until two years after the end of the war. My mother, Fridel Muehleisen Koesters, had to fend for herself and for my older brother, Jan; younger sister, Hanne; and me.

In the fall of 1944, I started grade school, but school did not last long. A bomb destroyed part of the building, and I marveled at the hole precisely where my classroom had been. I can't deny that I was pleased. That was the end of school for a while.

Our family lived in an apartment in a large old house with a sturdy basement of vaulted arches. There we spent days and nights during the waning days of the war, when Weimar was bombed by Allied forces. The sky was silver with planes, flying high. I remember well the wailing of the sirens, often at night, and the rush to get into the basement. People sat in chairs and on stools along the walls, cowering anxiously. They chatted about the anticipated horrors: bombs plastering you to the walls, fire engulfing you with no way to escape. I was terrified.

Then a mega bomb fell next to the house, across the street, into the soft earth of a park, forging a huge crater. It is dark in the basement. The planes come droning in, a low-pitched menacing hum. You hear the whistle of the bomb. The whistle gets louder and louder, nearer and nearer. You duck, you cover your head, you know this is the end. Then the explosion, the shaking, and quiet. The house was severely damaged, but the structure stood. We were to live another day.

The war was nearing its end. The American army was approaching from the west, the Russian army from the east. Weimar sits between them, with the infamous Buchenwald concentration camp only a few kilometers away. Rumors say the Allied forces will destroy the town in revenge for the camp. So, my mother fled to Bad Berka, a village twelve kilometers away. We walked, pulling a wooden cart that carried my little sister. The road was narrow and winding, traversing a low pass. Allied planes were strafing but held their fire because the road was filled with ragged, desperate-looking people, inmates of the Buchenwald camp. They marched out of the camp because the American army approached from one side, then marched back into the camp because the Russian army approached from the other side. My brother, one year older, remembers seeing people getting shot.

In Bad Berka, artillery shells whistled overhead all night. Next morning it was quiet, and I was sent to fetch some milk. Off I trudged a block or so to the store. The narrow road, flanked by stone walls, wound down a small hill. I rounded the corner and stumbled right into an oncoming American tank, filling the road wall to wall. The gun pointed at me; the white star glared at me. I ran.

Life After the War

My real life began on Armistice Day, May 8, 1945, also called VE Day, for Victory in Europe. It was a beautiful, warm, sunny spring day. I walked up an alley to visit a friend. The sky was blue, and I was going to live! No more sirens, no more bombs, no more days and nights in the cellar with my head under a mattress. No more fear of being buried underneath a house of rubble, being suffocated, plastered against the cellar walls, burned alive, dying. No more fear of dying. I was going to live.

But life had not yet returned to normal. We moved back to Weimar, into our apartment. It was a mess. American occupation forces, securing the place, had upended all the furniture, shot up pictures on the walls, and burned our building blocks. Yet we were lucky to be alive and have a roof over our heads.

For children, the American occupation was not bad. We ran alongside the tanks and yelled "Suugaar," and the GIs gave us their small gray C-ration cartons. I was so pleased that I made up my mind to go to America someday to partake in those riches. Little did I know this dream would come true. Six weeks after the war ended, the white-starred tanks left, and the red-starred tanks rolled in. Germany had been apportioned into occupation zones, and Weimar lay in the Soviet sector. My mother was furious. Why would the Americans give up territory they had won? She was also scared that the Russians would take revenge for what the Germans had done to them. Yet, for us kids, there was not much change. Now we had Misha instead of John. No more sugar, but the Russian soldiers overall were nice and played with us—except once, when an angry soldier stomped on our collection of snails. Why so vicious? Maybe somebody had killed his children.

We had little to eat but were used to it: two bare slices of rye bread per day, an evening meal of potatoes. My favorite dish was white sauce with little chunks of bacon on potatoes. That is all the meat we saw. We went into the country and foraged. We got skinny, but we did not starve. My bad teeth may have come from those malnourished days.

In 1946 rumor had it that the Iron Curtain, separating East Germany from the West, would come down. My mother decided to leave. At the border, an endless queue of people pushed and pulled wooden carts holding their belongings. Everybody was nervous and anxious, because the border might close any minute, and we would be stuck. I remember the tension and anguish. My little sister, a toddler of three, kept running away. We made it across. On the western side, we were doused with DDT disinfectant. My mother lied about my brother's age, because at nine years old, he would have been assigned to go with the men. You never, ever want to be separated from your children at a border, you may never see them again (sounds familiar?).

My aunt picked us up, and we made it to my paternal grandfather's house in Muenster. He was a doctor, with a white beard flowing in the wind when he rode his tall bicycle. The house, at the edge of town, was bursting with relatives crammed in because their houses had been destroyed. Muenster, an important railroad center, had been heavily bombed, and the entire center lay in ruins. Not until 1947 did we finally reunite with my father (figure 3.1), who had been a prisoner of war in England.

FIGURE 3.1 The Koesters family in 1947. *Left to right*, the author's older brother, Jan; younger sister, Hanne; a friend; the family dog, Bobbie; the author's father, Bernd; and the author. Courtesy of Baerbel Lucchitta.

A Budding Interest in Science

Life became peaceful, and I became curious. My siblings and I spent our summer vacations at my maternal grandmother's house in Swabia, near the Jura Mountains, which were topped by white limestone cliffs loaded with ammonite fossils. How, I wondered, did these sea creatures get on top of a mountain? When traveling to Swabia, the railroad went along the mighty river Rhine in a gorge cut into a mountain range. How could a river do this?, I pondered. My first scientific questions.

In 1949 I entered a Catholic girls' high school. There, I suffered through nine years of English, seven of French, and six of Latin, in addition to German, math, history, and natural sciences. The school was a forty-minute walk from our house. I walked or biked. In fact, the bicycle was my only means of transportation. My father's car was for his job, and my mother did not drive.

During my high-school years, I woke up to the magnificence of stars, my first foray into space. I admired Orion out my bedroom window. Then my physics teacher led excursions into the night, and I learned about the winter sky: Auriga, Taurus, Leo, and others. Sputnik 1 came in October 1957: beep, beep, beep. For the first time ever, humanity escaped the Earth. It was the beginning of the space age and of the space race between the United States and the Soviet Union. My brother and I were on a trip to Weimar, then in Soviet-controlled East Germany. The eerie sound and our somewhat precarious situation made for a memorable experience.

At the time I loved stories about travel and exploration. I read *Gods, Graves, and Scholars*, by C. W. Ceram, a book about the high adventure of archaeological discoveries.[1] For me this was a seminal book that awakened my interest in scientific research. In 1958 I finished high school and had to decide on a field of study. I still did not have a compass for my life. My mother pushed me to become a teacher because, if my future husband should die, I would have a job to fall back on. Woman were to get married; professions were just frills. But I was intent on a career. I did not want to be supported by parents or husband. I wanted to be free.

I was attracted to subjects ending in -*ology*: archaeology, anthropology, ethnology, and so on. I loved ornithology, an interest stimulated by a charismatic university professor, Ludwig Franzisket, a former World War II Luft-

---

1. C. W. Ceram, *Gods, Graves, and Scholars: The Story of Archaeology*, transl. E. B. Garside (New York: Knopf, 1951).

waffe fighter ace. He led early morning birding excursions. He was my first crush. Eventually I settled on geology because, contrary to my other *-ology* fantasies, it was more likely to lead to a profitable job. Yet, when I walked into the office of the University of Muenster's geology professor, Franz Lotze, he curtly dismissed me: geology is not for women. So, I put my desire on hold.

At the time, I had two girlfriends. Margaret Eickhoff was my travel companion on bicycle trips to Denmark and Holland. Ulla Roeren was my buddy on hitchhiking trips through France, England, and Scotland. Did we get accosted? Yes, but no harm was done. We told them to buzz off. We were expected to take care of ourselves: the Me Too movement, in which women have begun to publicly out sexual harassers, did not yet exist.

A New World

Then came the opportunity that changed my life forever: my childhood dream of going to America became real. I applied for a Fulbright scholarship for one year of study in the United States. I was the only female applicant. During my final interview by a committee of men, held in English, I was asked why I had listed my hometown when asked for my citizenship. I explained that *citizenship* obviously refers to "city." The men were so impressed by this unparalleled feat of logic that they gave me the grant.

So, in 1959, I was placed at Kent State University, considered a typical middle American school, where I soon realized that America was not the great land of freedom I had expected. I was beset by 10 p.m. curfews and dress codes. The college style was pleated skirts, Bermuda shorts, bobby socks, and sneakers, all of which I considered rather ugly. For the first year in the United States, I lived at Delta Gamma sorority in Kent, Ohio, which sponsored my room and board. I finally immersed myself in geology. In my classes I was usually the only woman. Indeed, being the only woman became the norm. It did not bother me; I was used to it.

After one year, my grant expired, but I stayed at Kent State for a second year, teaching German, to complete my bachelor's degree. A requirement was a geological field trip into the western United States. I was impressed by the endless prairies and wide open spaces. On the other hand, being used to the lovely green landscapes of Europe, the bare rock and dirt of some regions invoked images of gigantic construction sites, not perhaps the most flattering analogy.

During these years in the United States, I was by myself. I had neither relatives nor close friends whom I could ask for advice. My ambitions were not high; my aim was to stay in the United States. As my exchange-student visa could be renewed only as long as I went to school, I had to remain a student for as long as possible, which meant getting a PhD. This became my plan. After receiving my bachelor's degree in geology, I needed to decide on a graduate school, a choice that sets the course for the rest of one's life. I applied for assistantships and, as my grades were good, had several offers. I settled on the Pennsylvania State University in University Park (State College), Pennsylvania, which had a good reputation for earth science studies.

In 1962, at Penn State, I met Ivo Lucchitta, a fellow geology student and Czech-born immigrant from Italy. He remembers hearing me making my way to class, always late, with the clickety-clack of high-heeled shoes on tiled corridors; I had not yet shed my European ways. I specialized in structural geology because Robert Scholten, who had been a resistance fighter in Holland during the war, offered me a grant to study the Rocky Mountain thrust-fault belt on the Continental Divide in Idaho and Montana. I was reluctant because of my fear of bears, but Ivo convinced me to accept the grant. Even though I was the first female PhD in the geology department at Penn State, I do not recall any gender discrimination.

In April 1964 Ivo and I married because our landlady refused to continue tolerating our living in sin, and we didn't have time to look for alternate accommodations. It was a simple affair, involving a justice of the peace and a dinner with friends. Then the inevitable happened. I became pregnant. My well-laid plans were to complete my dissertation defense, then take two weeks off to study up on babies, and then deliver. As sometimes happens in life, these well-laid plans were thwarted. Baby Maya arrived the night before the scheduled exam in May 1966. The thesis committee was relieved, and my exam was postponed. I took it two weeks later, however, and passed without difficulty because I was still primed.

Fieldwork

Fieldwork in the wild west of the Rocky Mountains had its ups and downs, literally and figuratively. Being intent on getting a tan (nobody thought of skin cancer at the time), I wore short shorts and a tube top, accompanied by a leather belt with various pouches, heavy boots, and a hammer. I must have

FIGURE 3.2 Fieldwork in Montana. *Left*, the author and a colleague standing next to her ancient Jeep; *right*, sitting on camp tender Ralph Norquist's horse. Courtesy of Baerbel Lucchitta.

been quite a sight. Once, I snuck up on old sheepherder Joe, who almost had a heart attack when he saw this apparition. He quickly put on his hat to be properly dressed for the occasion.

A fellow graduate student, John M'Gonigle, procured an ancient jeep, and ranchers, shepherds, and cowboys helped me out by providing me with horses to ride, access to roads normally closed, and rescues from ditches when my jeep got stuck (figure 3.2). Such attention can backfire, though. One evening an irate woman rolled into camp with the intent of beating me up; I had allegedly seduced her husband. I had never met her, nor did I know the man; he must have fabricated the story to arouse her jealousy. Also, an enamored sheepherder threatened to shoot my then boyfriend Ivo when he came to visit. We moved across the Continental Divide as a precaution.

My field area comprised two-thirds of the fifteen-minute Morrison Lake quadrangle and straddled the Continental Divide. It was riddled with thrust faults. Here, I had my second encounter with space. On my very first day in the field, I described peculiar rock chips that had *conchoidal* (curved) fracture surfaces with fan-shaped grooves on them. I thought the grooves were strange fault striations. Later it became known that the chips were splinters of shatter cones. My former field area is now distinguished for harboring the largest impact structure in the United States, about 60 miles (100 km) in diameter. This was discovered only in 1990; in 1962, I had no idea what I was looking at. Indeed, only a handful of geologists at the time were even aware that impact craters existed on Earth, and I never noticed the spectacular outcrop of immense stacked cones nearby.[2]

Ivo's field area was at the mouth of the Grand Canyon, where he explored the history of the Colorado River. Because of the heat, he worked there in winter, whereas I worked in summer. This meant we had to travel back and forth across the continent several times a year. As I did not own a car, I traveled by Greyhound or Trailways buses. The first few hours of travel were long and tedious, then I settled into a kind of trance that lasted till the end of the trip.

The Sixties

Ivo and I both obtained our PhDs in 1966 and then faced the daunting challenge of finding employment for two. Ivo landed a job with the U.S. Geological Survey (USGS) Center for Astrogeology in Flagstaff, Arizona. It had two branches, Astrogeologic Studies, specializing in research and map making, and Surface Planetary Exploration (SPE), engaged in the practical aspects of training astronauts and mission development. Ivo worked for SPE, and I stayed home with baby Maya for one year.

Astrogeology was largely the invention of Eugene M. Shoemaker, a young, energetic, charismatic, and talented scientist, who was prevented by Addison's disease from going to the Moon. Instead, he settled in Flagstaff, then a small town of around twenty-five thousand people. It served as a proxy for the Moon, being close to the well-preserved impact structure Meteor Crater, as well as cinder cones and lava flows. Also, it had clear air for telescopic observations at Lowell Observatory. Not a minor consideration, Gene and his wife, Carolyn, liked to live there.

What was Flagstaff like in the sixties? It was connected to the world by the venerable highway US 66, going east-west, and US 89, going north-south. To go to Phoenix, one had to cross the mountains at Jerome. The interstates arrived only in the late sixties and early seventies. Manual typewriters reigned supreme. Duplication was by carbon paper and mimeograph. Computing was with IBM 360 mainframe computers, requiring punch cards and storage on magnetic tapes. Minor computing was done on a slide rule or mechan-

---

2. Regrettably, the cones were later vandalized by a local rancher, who probably sold them.

ical calculators. To communicate, one used rotary dial-up phones, walkie-talkies, and telegrams. Soon to come was faxing. Television was black and white. Personal desktop computers did not exist. Nobody thought of the internet, email, web surfing, Skyping, Zooming, mobile phones, smart phones, texting, or social media. But working as a professional had its perks: secretaries typed your manuscripts, photographers developed your pictures, and drafters made your maps and illustrated your images. There was only one building on McMillan Mesa, the one housing the Center for Astrogeology.

By 1967, I had grown bored with staying at home. The Astrogeologic Studies branch considered hiring me but could not because my exchange-student visa had expired. After a fair amount of paper shuffling, I obtained a green card that allowed me to work. I started out as a part-time technician. I didn't mind, as it gave me time to continue to tend to baby Maya. Finally, in 1972, I became a naturalized citizen and a full-time geologist with the USGS.

Mapping the Moon

At last, I had found my calling and a job I enjoyed. Thus began my career of looking at planetary images, which lasted for the rest of my working life.

First, I made geological maps of the Moon. Don E. Wilhelms, who was the foremost geological mapper at the time, taught me the fine points of mapping other planets and steadfastly supported even some of my more outrageous ideas. As early as 1960, when he was still in Menlo Park, California, Gene Shoemaker and his USGS colleague Robert J. Hackman had published a seminal paper on lunar stratigraphy, which conceived of lunar geological time in terms of five main periods (from youngest to oldest: Copernican, Eratosthenian, Procellariam, Imbrian, and pre-Imbrian), and this basic scheme was adopted and used for most subsequent lunar geological maps.[3] In the beginning, mappers of course still relied on telescopic observations. When I started mapping in 1967, the Lunar Orbiter 4 and 5 images had become available. They were analog photographs on film, developed on the spacecraft, then partitioned into strips, transmitted to Earth, and reassembled there. Thus, the images had stripes, each of which looked like a

---

3. See, for instance, Eugene M. Shoemaker and Robert J. Hackman, "Stratigraphic Basis for a Lunar Time Scale," in *The Moon: Proceedings of Symposium No. 14 of the International Astronomical Union held at Pulkovo Observatory, near Leningrad, December 1960* (London: Academic Press, 1961), 289–300.

step on a stereoscopic image. Later, during the Apollo missions, the geodetically controlled metric camera and elongated panoramic camera (PanCam) images provided stereoscopic viewing and fantastic resolution as small as two meters per pixel.

The lunar images, assembled into mosaics, were not ideal as a mapping base. They were illuminated by different Sun angles and azimuths and had many defects. So artists were recruited to produce controlled airbrush maps and to smooth out the imperfections. The main artists were Jay Inge and Patricia Bridges, both of Lowell and the USGS. I did not like the airbrush maps because it was difficult to transfer contact lines drawn on the images to the airbrush base. In addition, we inked the contact lines onto translucent plastic overlays, which was arduous. Eventually all lines had to be redrawn by skilled draftsmen. Digital methods were not yet used.

My first major assignment was to map the north polar area of the Moon.[4] It entailed mapping many craters and drawing a lot of circles (figure 3.3a). But I soon realized that the Moon offered more than craters; there were basins, mare, highlands, scarps, wrinkle ridges, rays, and dark mantles. I was assigned several detailed maps of potential landing sites for the upcoming Apollo missions. One, near the crater Proclus, had many small dark-haloed craters, then considered to be volcanic cinder cones. I quickly realized that they were impact craters excavating underlying dark layers. No volcanoes; too bad.

Apollo

Everybody who was alive and old enough to retain memories remembers where they were July 20, 1969, during the Apollo 11 mission, when Neil Armstrong stepped onto the surface of the Moon. I was in the middle of nowhere. My colleague John M'Gonigle and I were returning from a field excursion in Montana. Eager to get home, we took a dirt road shortcut, which was a big mistake. A violent thunderstorm poured heavy rain onto the road, and we were trapped between the raging Paria River and a landslide. Backtracking, we forded the flooding Paria and bounced through a soup of clay until our truck hit a submerged channel and drowned. We made it out of the car. A

---

4. B. K. Lucchitta, *Geologic Map of the North Side of the Moon*, U.S. Geological Survey Miscellaneous Investigations Map I-1062, scale 1:5,000,000 (Reston, Va.: U.S. Geological Survey, 1978).

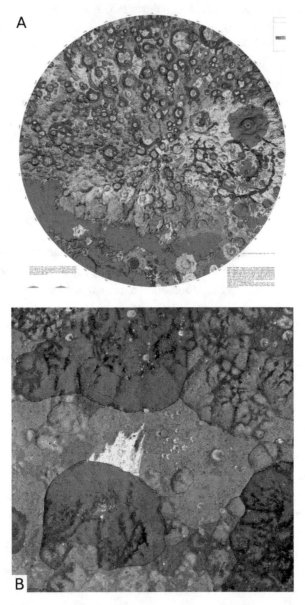

FIGURE 3.3 Geological maps of the Moon: (a) the north side of the Moon, scale 1:5,000,000; (b) the Taurus Littrow region (Apollo 17 landing site), scale 1:50,000. Courtesy of Baerbel Lucchitta.

local rancher took one look at us and said, "Yep, first storm of the season, happens all the time." He gave us a ride to Ruby's Inn, at Bryce Canyon National Park. There, we watched the first blurry steps onto the Moon on the only TV, in the lobby. At 10 p.m., the owner-lady turned off the TV: bedtime. No arguments accepted. The truck was later retrieved by a USGS crew.

My breakthrough came when I prepared the 1:50,000 scale geological map of the Apollo 17 landing site (figure 3.3b). It was one of only a few planetary maps checked in the field. Up to that time, as a young woman, I was considered a junior person. Here was my opportunity. When I attended a meeting about prospective Apollo 17 landing sites (figure 3.4), I noticed that Harrison H. (Jack) Schmitt, the geologist astronaut for this mission, was seriously interested in the Taurus Littrow region, perched on the southeastern rim of the Serenitatis basin. I quickly asked my supervisor to change my assignment and let me map this area, and he concurred. The site was indeed selected. My intuition proved me right.

During my subsequent careful mapping of this site, I realized that the highlands were made of breccia. At the time, it was still widely believed that they were composed of light-colored silicic volcanic material. Apollo 16 was sent to the highlands to clarify this issue. I tried to convince the Apollo 16 crew, John Young and Charlie Duke, before their flight that they would find

FIGURE 3.4 The Apollo days: *Left*, Apollo 17 landing site-selection meeting, with, *from left to right*, Gordon Swann, R. Henry (seated), Hal Masursky, Jack Schmitt, Hal Stevenson, and the author standing modestly in the background on the right; *right*, the author posing on Grover, the USGS's simulated lunar rover, used by the astronauts to prepare for lunar roving vehicle excursions on the lunar surface, carried out on the Apollo 15, 16, and 17 missions. Courtesy of Baerbel Lucchitta.

no volcanic rock, only impact breccia. They did not believe me. Nobody believed me. Yet, I was right.

I flew to Cape Canaveral twice to give lectures on the geology of the Apollo 17 landing site. On the first trip, the plane was late, so I would be late for dinner at the astronauts' quarters. Then, I had a flat tire. Pressed for time, I moved on despite the flat. At the dinner, Ron Evans, the command module pilot for Apollo 17, was late as well. When he entered the room, he looked around the table and said, "Where is Dr. Lucchitta? He is supposed to give us a lecture tonight." All fingers pointed at me: "It is *she!*" I suppose he did not expect a young woman. After my evening lecture, Ron Evans looked at my shredded tire, declared that it was not fixable, and gave me a ride first to a bar to have a drink, and then to my hotel. I informed the rental car company the next day, and they retrieved the car.

During the Apollo 17 mission, December 7 to 19, 1972, I was in Houston with the backroom support team. We transmitted our findings and instructions to the command center, which communicated with the crew on the Moon. I was eager to find out about the accuracy of my geological map. Overall, my interpretations were on target, but I had interpreted a dark mantle, apparently covering the area, to be young because it showed few craters. It turned out to be old; it was a thick regolith (lunar soil) in which young craters were readily eroded and thus disappeared. More importantly, I had proposed that orange material of volcanic origin might be found at the site, because I had noticed that elsewhere it was associated with dark mantles. Orange material was indeed found. Jack Schmitt and I published a paper on this topic. In the meantime, I received some notoriety in a German tabloid, which published a picture of me with the caption, "This woman lets the astronauts dance to her tune." I was furious.

Mapping Mars

Apollo 17 was the last trip to the Moon. The public lost interest, the space race had been won, and funding dried up. Interest turned to Mars. In January 1972, the first planetary orbiter, Mariner 9, furnished comprehensive views of Mars after the dust from a planetwide storm finally cleared. The images revealed many Earth-like features, such as volcanoes, possible riverbeds, depositional mounds, and the 2,500-kilometer-long canyon Valles Marineris, named after this mission. I was an unofficial adviser to this mission and happy to be part of the action.

After the mission, Stephen Dwornik, of NASA Headquarters, set up a formal program of geological mapping of Mars quadrangles at 1:5,000,000 scale. I mapped the Ismenius Lacus Quadrangle, positioned along the northern boundary of the highlands, which here were deeply dissected by canyons with grooved floors. I inferred glacier-like processes for their origin, my first foray into ice on Mars. In 1976–77 the Viking spacecraft arrived on Mars, and the improved images brought many geomorphological features into better focus. This time, I was a formal guest investigator on the Viking team. I was thrilled when the lander images built up on the screen, line by line, showing the ground on Mars for the first time ever.

Then came a great hiatus in Mars exploration. Funding dried up, and no missions went to Mars until 1996. For those twenty years, most photogeological research was based on Viking images. Ivo and I took the opportunity to go on private trips. Besides visits to Europe, where my parents still lived, we hiked in the Wind River Mountains of Wyoming, rafted the Grand Canyon and many other smaller rivers in the West, and enjoyed road trips through the western United States in our trusted VW Westphalia camper. Places were not as crowded then as they are now. I also served as associate branch chief of astrogeology at USGS from 1986 to 1991.

The Jovian Satellites

In the meantime, images arrived from the Voyager 1 and 2 spacecraft, which flew by the moons of Jupiter in 1979 on the first leg of their epic exploration of the outer solar system. I shifted gears. I was at the Jet Propulsion Laboratory (JPL) when Voyager sent the first closeup views of Io, Europa, Ganymede, and Callisto. Oddly enough, in a paper published just three days before the Voyager 1 flyby in March, dynamicists Stanton Peale, Patrick Cassen, and Ray Reynolds had suggested that Io's interior might be molten, owing to tidal interactions with Europa and Ganymede. In their own words: "Calculations suggest that Io might currently be the most intensely heated terrestrial-type body in the solar system. . . . One might speculate that widespread and recurrent volcanism would occur, leading to extensive differentiation and outgassing."[5] Their idea was confirmed when Voyager sent back images of volcanic plumes rising above the limb of Io and showed that the surface was

---

5. S. J. Peale, P. Cassen, and R. T. Reynolds, "Melting of Io by Tidal Dissipation," *Science* 203 (1979): 892–94.

utterly devoid of impact craters. Io was casually dubbed "the pizza," because of its red and yellow color, punctuated by olive-like dark spots. Regrettably, we saw only a small part of Europa, which showed tantalizing cracks that looked like sea ice. We were surprised by Ganymede, the largest of Jupiter's moons, which showed sharply delineated regions of light, grooved terrain, and more heavily cratered dark terrain. Callisto was ubiquitously cratered. It became increasingly obvious that the bodies of the solar system are unexpectedly diverse.

After the missions, I, with other researchers, published the first geological maps of Europa and Ganymede. Already, we predicted the existence of water under the ice of Europa. From 1980 to 1995, I led the Galilean Satellite Geologic Mapping Program, with the help of my assistant Holly Ferguson. The program included forty researchers from eight nations, and leading it was like herding cats. From 1995 to 2003, the Galileo Jupiter orbiter filled image voids left by Voyager, thus causing a renewed geological mapping effort. I was not involved because I had moved on to other subjects. One might almost say I had come down to Earth again.

Ice in Antarctica—and on Mars

In the early eighties, my branch chief, Larry Soderblom, encouraged digital processing of Landsat 1 satellite images of Antarctica, acquired in 1972. They were available only as poorly developed, underappreciated, black and white paper prints. I managed this massive effort with the help of image-processing expert JoAnn Isbrecht. What I saw deepened my interest in ice-related features, and I realized that similar landscapes were seen on Mars. I then honed my glaciology skills with a field trip to Siberia in 1982. My luggage never arrived, and I spent ten days on the Lena River looking at ice wedges and frozen polygonal ground, all the while wearing the same jeans. My Russian colleagues wondered whether I could not afford a change of clothes.

I lucked into a trip to the Northwest Territory of Canada when a colleague became ill and dropped out. With the help of John Behrend of the USGS in Denver, I went to Antarctica in 1984 to learn about glacial and periglacial processes. I had a suitcase full of highly desired, colorful Landsat images portraying coastlines and glaciers, which I handed out freely. Thus, I bribed my way onto helicopter rides to the Dry Valleys, blue-ice meteorite collection sites, and the famed Lake Vanda, where a swim in the ice-covered water was de rigueur. I also hopped on a plane to the South Pole and brought home many blurry pictures because my camera fogged up.

The early satellite images, when compared with later ones, showed the shrinking or expansion of ice shelves and enabled the tracking of crevasses, which then yielded glacier velocities. Using this method, Ferguson and I measured the speed of the giant Byrd Glacier, and Christie Rosanova and I measured the speeds of the giant Thwaites and Pine Island Glaciers and of lesser glaciers along the West Antarctic coastline.

My newly found enthusiasm for ice-related landscapes led me to look for them on Mars; I was astonished to find many. Debris flows in the Valles Marineris and elsewhere looked like rock glaciers in Alaska. Polygonal ground in front of outflow channels suggested mudflats from a dried-up ocean. Collapsed ground resembled thermokarst. Parallel grooves and ridges on the floors of outflow channels looked like those on the floors of terrestrial glacial valleys and ice streams, which led me to propose that ice sculpture and not floods gave these channels their characteristic appearance. I tried to convince my colleagues that ancient Mars had not only been wet, but had, and might still have, massive reservoirs of ground ice. My papers, though they were to prove prescient, were ignored at the time.[6]

Shifting away from ice, I concentrated on the Valles Marineris, whose origin is still debated. Are the "valles" of tectonic or erosional origin? Did they ever harbor lakes and give rise to rivers? How did the sediment piles in its interior build up? I studied landslides, prepared a geological map, and published a major paper summarizing the arguments for and against lakes. The paper appeared in a book, *Lakes on Mars*, but is mostly forgotten today.[7]

Resuming Space Exploration

After the long post-Viking hiatus, Mars missions finally resumed, and one followed another in rapid succession, furnishing a cornucopia of images and other data. In 1996 the Mars orbiter camera's black and white images showed never-before-seen detail on the ground. A laser altimeter on the same mission for the first time furnished accurate topography. In 2001 the

---

6. B. K. Lucchitta, "Mars and Earth: Comparison of Cold-Climate Features," *Icarus* 45, no. 2. (1981): 264–303.

7. B. K. Lucchitta, "Lakes in Valles Marineris," in *Lakes on Mars*, edited by N. A. Cabrol and E. A. Grin, 111–61 (Dordrecht: Elsevier, 2010).

thermal emission imaging system stood out because of its many spectral bands. The German Mars Express orbiter flew in 2003. The pictures of the High-Resolution Imaging Science Experiment (HiRISE) in 2005 had an astonishing resolution of 0.3 meters per pixel in selected areas. This mission also covered most of the planet in black and white with the context camera (CTX); also on board, the compact reconnaissance imaging spectrometer (CRISM) delivered visible, near-infrared, and infrared images. The MAVEN orbiter flew in 2013 and the ExoMars orbiter in 2016. The Mars Pathfinder with Sojourner rover landed in 1997, the Spirit and Opportunity rovers in 2003, the Curiosity rover in 2011, the Insight lander in 2018, and the Perseverance lander and Ingenuity helicopter in 2021.

Mars is now much better understood, and for me personally, the most gratifying observations involve ice on the planet. Recently, outcrops on young scarps on Mars have been found with near-surface water ice more than one hundred meters thick, perhaps deposited during high-obliquity periods.[8] It is commonly accepted that water ice indeed played an important role in the geomorphic evolution of Mars.

FIGURE 3.5 The author celebrating her eightieth birthday with a cake showing the sun and planets. Courtesy of Baerbel Lucchitta.

A Few Final Thoughts

In summary, my two guiding principles "go with the flow and grab the opportunities" and "keep an open mind about alternative ideas" have served me well. They brought me to the United States; led me to a PhD; enabled me to meet my husband, Ivo Lucchitta; and eventually set me on a path to planetary geology and Mars. They served me well enough to reap some accolades, among them a Meritorious Service Award from the Department of the Interior in 1994; the Gilbert Award of the Plane-

8. C. M. Dundas et al., "Exposed Subsurface Ice Sheets in the Martian Mid-latitudes," *Science* 359, no. 6372 (2019): 199–201.

tary Geology Division of the Geological Society of America in 1995; a glacier named Lucchitta, awarded by the Advisory Committee on Antarctic Names in January 2003; and an asteroid named 4569 Baerbel, discovered in 1985 by Gene and Carolyn Shoemaker. I retired in 1995, but now, over eighty years old (figure 3.5), I still enjoy keeping up with the latest developments in planetary science. May the future hold many more missions to Mars, including manned landings. There is so much more to learn.

First Light

Personal Reflections on Exploring the Moon and Planets

PETER H. SCHULTZ

When I was in the seventh grade, I submitted a photo of comet Arend-Roland to *Sky & Telescope* magazine in hopes of having it published (figure 4.1). Flash-forward sixty-five years. I'm walking across the Atacama Desert in Chile trying to explain what produced widespread fields of glassy slabs scattered across the ground. Both events are connected across time. My boyhood passion for astronomy and my fascination with geology led to my career in planetary geology, researching surface processes on the Moon, performing theoretical and experimental studies of impact cratering, and serving as a co-investigator on NASA missions to two comets (Deep Impact and Stardust-NExT). This is the story of how I ended up in such a far-flung place on a journey of discovery that still continues to this day.

I come from a family with diverse backgrounds and interests. My father was a biochemist who became director of research at a major pharmaceutical company. My mother was an accomplished portrait artist who studied in Florence and was one of the first women to enter Yale's graduate program in fine arts. My older brother became a well-known and respected dentist in Lubbock, Texas, while my younger sister dealt with crisis management (I'm sure because of her two brothers), rising to vice president of a major company. Like most growing families during and after World War II, our needs were simple: small postwar homes on narrow tree-lined streets, in our case in the Midwest (Michigan and Indiana). Normal life had been on hold during WWII and the Korean War but rapidly changed with new opportunities.

FIGURE 4.1 Photograph of comet Arend-Roland taken by the author in April 1957, just after he turned thirteen years old. The broken streak below the comet is from a plane arriving at Chicago's O'Hare Airport, which had just opened for passenger travel two years earlier. Courtesy of Peter H. Schultz.

Within ten years, I would live in a middle-class suburban home far enough outside Chicago's lights to allow viewing under a dark sky. Just like many kids in third grade, this is when my interest in both astronomy and geology began (figure 4.2).

The seeds, however, were subliminally implanted earlier by the convergence of four experiences. First, I vividly recall a coloring book with rockets while still in preschool in Indiana in the forties. Second, I was given a small two-draw telescope while in second grade for my birthday. Although I don't recall asking for a telescope, this little metal instrument on a tiny tripod provided my first close look at birds, animals—and the Moon. Third, theaters started showing movies like *Commander Cody*, *It Came from Outer Space*, and Disney's *Conquest of Space*. The ultimate, though, was the classic *The Day the Earth Stood Still*. The fourth seed was the *Life* magazine series The World We Live In, featuring Chesley Bonestell's cover *The Earth is Born* (December 1952) and a foldout of a dinosaur mural (*Pageant of Life*, Peabody Museum, Yale) painted by Rudolph Zallinger. Such representations sparked my imagination of deep time. These four experiences did not chart my path,

FIGURE 4.2 The only way to land on the Moon in 1957 was to visit an amusement park. *From left to right*, the author's good friend, who would develop polio the next year (the Salk inactivated polio vaccine, or IPV, had been licensed only two years earlier); the author; and his sister and father. Courtesy of Peter H. Schultz.

but they did reveal how doors open if everything else aligns. How did such simple activities evolve into an obsession?

As a kid, I had severe asthma. I spent part of third grade in a hospital under an oxygen tent. Although I missed school, I decided to memorize the properties of all the planets in the solar system (size, distance, etc.). After getting out of the hospital, I took weekly hydrocortisone shots for years, along with the allergy tests and treatments. Synthetic cortisone had been developed less than a decade previously, and its long-term effects were not fully known. My prolonged use of high-dose cortisone injections produced a typical puffy appearance at that time (and may have stunted my growth). But at least I could walk (rather than crawl) up the stairs.

That time in the hospital forced me not only to *internalize* facts but also to imagine traveling to each object. During this time, my third-grade teacher (Miss Jackson) had an interest in geology and took the entire class to a coal-mining quarry in Coal City, Illinois, where we broke open rocks to reveal fossil ferns resembling those depicted in *Life* magazine. But I also found a rock with an impression of a trilobite, which produced a surge of the curiosity-related neurotransmitter dopamine. This *discovery* personalized science and made me *curious* about how my trilobite had lived, died, and eventually become trapped in the rock. I actually *owned* a once-living creature that had crawled across a Silurian marine bed. This had to be shared, so I made several plaster casts, which I colored to look just like the fossils I had seen in museums. While my friends had lemonade stands, I tried to sell my fossil casts from my red Radio Flyer wagon. Aside from a single pity purchase by a neighbor, I quickly learned a lesson: not everyone shares my enthusiasm. It was also a poorly conceived business model. These four key ingredients (internalization, discovery, curiosity, and ownership) would eventually shape not only my research but also my approach to teaching.

The northern Illinois plains occasionally have large glacial erratics—boulders dropped by the glaciers after retreating. A friend of mine and I discovered one such three-foot boulder lying alone in a field, avoided by the plow. It was filled with crinoid stem fossils (a.k.a. sea lilies) dating back about four hundred million years. Every day we rode our bikes to this treasure until there was nothing left but rubble. One the one hand, we had our prey; on the other, we had destroyed its time capsule. Then we saw the remarkable diorama of a Paleozoic reef at the Field Museum in Chicago.

In third grade, one of the ways to further develop these ingredients was to draw what I saw. I mentioned the dinosaur mural in *Life* magazine and my mother the artist. These two anecdotes actually connect and become personal because my mother went to Yale while Zallinger was there. She passed on the "art gene," which I exercised by making freehand copies of the Peabody murals. Sketching the dinosaurs in the mural required more than just copying; I had to interpret joints, muscles, colors, and textures. I even asked my parents for a human skeleton for my birthday but only got a roll-down mural of a skeleton and the musculature. Many years later, I explained to my father that I didn't go into medicine because I never got my skeleton. I was kidding, of course (hope he got the joke). Regardless, my father instilled curiosity; my mother, my ability to observe.

My childhood was pretty normal because of my brother and sister, but in retrospect, I was different. I created puppet shows for the neighborhood; made ice sculptures of Abe Lincoln; re-created the layout of O'Hare Airport with string in the backyard; made space ships out of refrigerator boxes and upside-down fishbowls; saved toilet-paper rolls for my Commander Cody jet pack; and played with a monorail on a wire instead of a Lionel train on tracks. I also developed an obsession with farm silos (perfect for an observatory).

By the fifth grade, I wanted a larger telescope, and my parents bought me a 4-inch Dynascope reflecting telescope, manufactured by the Criterion Company of Hartford, Connecticut. With my face pressed against the Bakelite tube, I inhaled its distinctive smell, which I still recall. My first light came from the first quarter Moon. This virtual image of the Moon was completely unlike the photograph from Lick Observatory pinned above my bed. Seeing the Moon go directly into my eye through a telescope was like being there. I had to see for myself each phase, from the nascent crescent before sunset to waxing sliver before sunrise, and decided to make my own miniature Moon. I shaped a hemisphere out of clay and kept it wet with a towel over successive nights while trying to reproduce what I saw. This forced me to focus on each feature. I used flathead nails for the large craters and finishing nails for the smaller ones. I would learn later that I was capturing complex and simple craters (figure 4.3). Now I had my own personal Moon.

While I continued to draw and paint, I felt the need to capture (and own) what I saw through the telescope. This led to taking photographs of the Moon, planets, and the Sun with my father's Argus C-3 (known as "The Brick") adapted to my telescope. After hundreds of nights spent using different types of film, I developed and printed the results in the back of my shared closet with different techniques, such as stacking negatives. I then started making my own craters by blowing up ant mounds. Using my Brownie Hawkeye

FIGURE 4.3 Over many nights, the author sculpted a clay model of the Moon as seen through his new 4-inch Dynascope reflector. A wet towel kept the model moist over successive nights. Courtesy of Peter H. Schultz.

camera, I photographed ballistically launched ants, never knowing that years later I would use high-speed cameras to capture impact ejecta up to two million frames per second.

The stars seemed to align in 1957, with the International Geophysical Year (IGY), the solar maximum, two bright comets (Arend-Roland and Mrkos), a transit of Mercury, and the launch of Sputnik. Although many activities surrounded the IGY, I was obsessed with photographing sunspots and magnificent aurora borealis displays, which looked like cosmic finger painting. My first published photograph captured the aurora reaching northern Illinois and appeared in *Sky & Telescope* that year. I also tried to submit my old photo of Arend-Roland, but there was a lot of competition; in any case, it was never published.

On one rather memorable occasion, my telescope almost blinded me and came close to getting me arrested. A transit of Mercury occurred on May 6, 1957; while watching it, the glass filter at the eyepiece cracked. Just when I heard a pop, I moved away, barely avoiding serious injury. And I was nearly arrested while trying to see the green flash with my telescope. The *green flash* is an atmospheric phenomenon that occurs just at sunset (or sunrise) under extremely clear conditions. As the Sun disappears below the horizon, the top of the Sun appears to detach suddenly and create a red (or less commonly, a green) flash. I focused in as the Sun approached the horizon but didn't see military police approaching me. Between the Sun and this bazooka-like telescope was a Nike-Hercules antiballistic missile site. In the fifties, Nike missile sites ringed large cities like Chicago, and military police constantly monitored their perimeters—and apparently me. I didn't get to see the green flash. But fifty years later, conditions were perfect as I watched the Sun set over the Pacific horizon. As I turned to tell my wife what to look at, she saw the green flash. I still haven't seen it.

Watching Sputnik 1 and 2 cross the night sky turned science fiction movies into a new reality. The satellites were just barely seen with the naked eye, but the nightly sightings actually normalized me in the eyes of friends. They called me Satellite Sam because I was small enough to fit in a space capsule. Astronomy now provided a context for all my math and physics classes, while photography provided context for chemistry. I had a purpose, not for a career but just to understand. Landing on the Moon or visiting every planet in the solar system could only be imagined. Within a decade I would become part of the first generation of planetary explorers.

At about the same time, I was entranced by physicist Daniel Q. Posin (DePaul University), who lectured on astronomy and astrophysics on WTTW-TV (Chicago), even though the physics was over my head at the time. Although he was rather dry and spoke with a Russian accent, I found him highly engaging. He wasn't afraid to mix things up a bit by throwing an eraser at his cat, Midnight, or to digress in the interest of telling a good story. Without knowing it, I actually adopted some of his teaching style—including my share of thrown erasers as well as random anecdotes to personalize science. His example stimulated my participation in numerous television science programs over the last thirty-five years. Even though I knew I wouldn't hear from one of the kids who were tuned in, I knew they were out there based on chats with mothers, occasional strangers, and bus drivers.

In Arlington, my asthma prevented me from playing "normal" sports, and I had to take a "special gym class" with students with various physical (e.g., giantism, polio, cerebral palsy) and learning disabilities. One of these students became a fellow astronomy buddy. This experience taught me to accept people as they are and value their uniqueness. As the severity of my asthma waned (though still chronic), I was able to take up less strenuous sports, such as bowling, fencing, lifting weights, and golf; my siblings kept me from disappearing into a book. Other than my asthma and my obsession with astronomy, I had a typical fifties student life in Illinois: dances, football and basketball games, and parades. As a sophomore, I even designed and helped to build the prizewinning homecoming float.

My family moved from Illinois to Lincoln, Nebraska, in November of my junior year. I heard about a fencing club at the University of Nebraska, but when I arrived at the meeting time, there was a judo club instead (the fencing club had disbanded). So I took up judo, which would become an important part of my recovery from asthma over the next four years. I even gave up golf (cold turkey) because it took up too much time, and I became too intense. Once on the third hole, I imagined that if I didn't make the put, the universe would implode. I missed and then walked away to prepare for the end of time.

Just before moving to Lincoln, I had started grinding my own mirror blank for a new telescope but ran into a large bubble on the surface. In truth, I was more interested in looking through the telescope than grinding the lens. Instead, I constructed a 10-inch telescope with a long stainless steel shaft (for the equatorial axis) that I put inside a friction plate inside a cast-iron drainpipe. Rather than using an expensive motor and gear for the clock drive, I made an extended guillotine-shaped arm, where a slightly bent rack-and-pinion system was attached to a small variable-speed motor. This design allowed tracking stars (or the Moon) at low cost for a couple of hours before resetting (figure 4.4). While passing by a pawn shop in Lincoln, I saw several giant brass lenses on the top shelf, once used in a local photo studio but seemingly perfect for taking wide-field astrophotography. The pawnbroker told me to get lost. Instead, I bought a WWII surplus f/2.5 7-inch Aero-Ektar lens. I would later learn that a similar large brass lens had been used on the Crocker Astrograph at Lick Observatory. Fifty years later, my wife and I had collected hundreds of those big brass portrait lenses.

While in high school, I tried to get a job at NASA's new Prairie Network Program, which was based in Lincoln starting in 1962. Run by Harvard's Smithsonian Astrophysical Observatory, this program used surplus wide-angle aerial cameras at sixteen stations across the Midwest to monitor meteors, determine preentry orbit, and hopefully recover the meteorites. Outside I saw a black-painted milk-delivery truck, with a flaming meteor on the side,

FIGURE 4.4 While in high school, the author designed and built this 10-inch reflector that would allow tracking the stars for about three hours. Variable-speed motors enabled adjusting for the Moon's motion. Courtesy of Peter H. Schultz.

that was used to carry new film to each site and return the exposed film to the base. This seemed perfect for me: I could chase meteorite falls, exchange film, or just hang around. After entering a giant room full of wide-angle lenses, oddly shaped cameras, and equipment, I asked if I could work there over the summer. They were kind, but internships were not popular then.

There's another reason I mention this. I would learn later that my graduate adviser had worked on a project at Harvard counting meteors while sitting in an old barber chair on the back of a truck. This might explain why years later, my wife and I refurbished antique wooden barber chairs for relaxing and watching TV. By the way, I actually ran across one of those Prairie Network chase trucks with the flaming meteor painted on its side for sale at an outside antique market in New England. I was giddy and tempted; my wife, not so much.

My experience with astrophotography led to summer jobs. In Illinois I worked at a commercial studio, where I printed enlargements of industrial tools and compiled them into a handbook. But occasionally, I was at the front counter selling film and cameras. This is where I saw my first antique wood and brass camera with long red leather bellows. It belonged to a customer who had loaned it for display, and after two months of ogling, I finally was able to purchase the camera for twenty-five dollars (a lot of money back then). I used it to make a fake ID for my older brother in exchange for a golf bag (fortunately, I believe the statute of limitations has now passed). It was also the start of another obsession shared with my future wife: collecting antique cameras.

In Lincoln, I worked in a film lab called the Film Shop. Now my job involved driving a jeep around town to pick up film for processing from different pharmacies, and then return the finished products that afternoon. Back in the shop, I helped to process film using an automated developing system, which had been designed by the owner. The system carried hanging film strips clipped to a moving frame that lifted from one chemical bath and then lowered to the next. Sawlike teeth made sure that the hangers rested in the correct bath for the allotted time. Occasionally the film strip would release from the clip and fall to the bottom of one of the baths, which could ruin the negative. The rule was to just let it go. Any attempt to retrieve the film could get you guillotined, even though it was supposed to have a braking system if it felt any resistance. But when it happened to me, I just couldn't leave the film on the bottom of the vat. I felt the armature coming down on the top of my head, turned my head, and felt the teeth scrape across my cheeks before barely escaping.

My summer job over the next several years was less dangerous: selling cameras in a department store. Such experiences brought balance: an introvert focused on astronomy and an extrovert talking to customers. Nevertheless, the manager reprimanded me for disappearing into the backroom over lunch to read *Sky & Telescope* or *Scientific American*. He thought this was not healthy for a teenager. So, I reluctantly did what he asked and wandered aimlessly around town until I found another place to hide and read. I also would take cameras home over the weekend to test them out—sometimes for astrophotography and other times for taking portraits. One of the cameras was a gorgeous old German folding camera. Sixty years later, I was having a casual conversation with a stranger who said he had lived in Lincoln in the early sixties when I was working in the camera store. Turns out he had bought this very same camera.

While I was doing the twist at a sock hop after a high-school basketball game in Lincoln, this cute girl cut in during a Sadie Hawkins dance set. All through my senior year, we were inseparable, whether cruising the carhop (imagine the movie *American Graffiti*), going to drive-in movies, or dancing at sock hops. We also both enjoyed art, so we would go into the Nebraska countryside to paint and sketch old barns, animals, and houses. After six months of dating, I asked her mother if she could stay up all night to count meteors during the Perseid shower. I must have seemed harmless (actually I was) because her mother consented. We stayed out in the middle of a friend's farm near a small town south of Lincoln and tallied fifty meteors per hour. In retrospect, I guess it was a great line: we've been together for more than sixty-one years.

Our meeting back then may have been part of some larger cosmic plan. My wife was adopted soon after birth in Portland, Oregon. After moving many times, she and her adopted parents ended up in Lincoln, just a few years before I arrived. While doing a genealogical search fifty years later, we discovered that her birth mother and family had not just lived in Nebraska in the forties but had come from the very same small town where we had watched the meteor shower. Coincidence?

While in high school in Lincoln, I cofounded the Prairie Astronomy Club and was its secretary for several years. The club is still going strong sixty-one years later, with active members, a dedicated observatory, and great newsletters (figure 4.5). I evolved two lives. One was the geek who enjoyed staying up all night with my geek friends, talking astronomy, comparing telescopes, or painting. The other life was an outgoing kid doing things with his crazy high-school friends at football or basketball games and playing crazy pranks at the Nebraska State Fair. One time, the two lives merged. I decided to make

FIGURE 4.5 Using a flashbulb, the author made three exposures of himself (arrows) and his astronomy club friends in different positions around their telescopes. The author's homemade astrograph is at the center, while his 10-inch telescope is at right. The photo was taken on a farm owned by a fellow amateur in Hickmann, Nebraska (south of Lincoln). This is also where the author and his girlfriend used to stay up all night counting the Perseid meteors. Courtesy of Peter H. Schultz.

caricatures of all my teachers. Early one morning I taped them to the wall in my high-school hallway entrance, sort of a Rogues Gallery for my classmates. Fifty-six years later, I was humbled to receive a Distinguished Alumni Award from my high school in Lincoln. At the award dinner, I was shocked to see one of my subjects, my ninety-two-year-old math teacher, sitting at our table. He then told me that students had rushed into his classroom that morning, saying that he couldn't let me back in class after what I did to his face. At the dinner, he also told me that he had kept my caricature on the wall in his office until passing it on to his son, who still has it hanging in his office.

After high school, I attended Carleton, a small highly regarded liberal arts college in Northfield, Minnesota. I chose this college because it had a historic observatory (now listed on the National Register of Historic Places). Its first iteration had been built in 1878—two years after the famous raid on the Northfield Bank by Jesse and Frank James and the Younger brothers— but the red-brick building that now exists was built in 1887 to house two telescopes—an 8.25-inch Clark and a 16.2-inch Brashear refractor. In the observatory's heyday, it had also been the headquarters from which Carleton faculty member W. W. Payne operated the time service that kept the trains of the midwestern prairies running on time and edited the journals *Sidereal Messenger, Popular Astronomy,* and *Astronomy and Astro-Physics* (the last co-edited with George Ellery Hale and the precursor of the *Astrophysical Journal*). At least as important was the fact that in Minnesota, I was closer to the northern lights than in Nebraska or colleges back east. Of course I asked if I could have a bedroom with a northern exposure so that I could watch the aurora. There were two issues with this request. First, my room had a giant tree in front of the window, which blocked my view; second, my leaky window was exposed to the long, windy Minnesota winters.

While Carleton College was a perfect fit for this midwestern kid, there were also adjustments. My roommate was from New York City (I had never met this particular species before), and so many other students were much smarter than I was. I learned to accept and appreciate both awakenings. I was overly active in my first year of college: I wrestled, was voted class VP, and took all the required first-year courses. While taking introductory physics, chemistry, and math classes, I invariably found myself translating assignments into questions related to astronomy. In other words, context and relevance gave more meaning to the problem sets. And I took studio art classes, which resulted in a battle between the left and right sides of my brain. I continued to make caricatures of friends; however, prudence dictated that it was best to avoid drawing faculty.

After recalling my fascination with dinosaurs and evolving landscapes, I thought it would be fun to take an introductory class in geology. This class rekindled my love of looking back in near time (i.e., Earth) rather than deep time (i.e., astronomy). On airplane trips, I would carry one of the texts about the regional geomorphology of the United States, something I would do well into the seventies. Back then I could send notes to the pilot as we flew over Mars-like landscapes, and he would announce it to the cabin (can't do that anymore).

For my senior honors thesis, I decided to merge my interest in astronomy and geology by diving into the debate about the origin of lunar craters. The first detailed images of the Moon had yet to be returned (except those from the kamikaze Ranger series). The astronomy professor allowed it, and the topic drew me into a fascinating trip around another world, with its scientific debates, contrasting disciplines, and strong personalities. About thirty years later, I returned to a similar theme during my first sabbatical on the faculty at Brown. My paper "Shooting the Moon: Understanding Lunar Impact Theories" focused on why it took so long to acknowledge the origin by impacts.[1] After all, telescopes were large enough, and the basic physics was known. The answer was a combination of visualization, warring disciplines, and science politics.

The classic refractors housed in the historic observatory building had been legitimate research telescopes in their day, and they recalled the exciting era when astronomers argued about the existence of canals on Mars and were just becoming aware of the colorful and ever-changing phenomena in the clouds of Jupiter, including the famous Great Red Spot, which first came to prominence about the time of the observatory's founding. Properly vetted students were allowed access to the instruments. The large dome housing the 16.2-inch Brashear had to be moved by pulling a rope attached to a gear. I had to literally jump up, hang on the rope, and brace myself against the wall to move the dome before climbing a tall ladder to look through the telescope. Now this was astronomy! Majors took spherical and practical astronomy and were trained on a large (5-inch) Repsold meridian circle, which Payne and his colleagues had once used for the timekeeping service. This experience led me (as well as many other Carleton astronomy majors) to work at the Nautical Almanac Office at the U.S. Naval Observatory in Washington, D.C., over the summer of my senior year (and the year following). Even though cesium clocks had replaced timekeeping using meridian circles, *The American Ephemeris and Nautical Almanac* (known since 1981 as *The Astronomical Almanac*) still provided the ephemerides (precise predicted positions) of celestial bodies. Interns at the office assisted in these calculations, as well as in related research. On the Naval Observatory's campus was the 26-inch Clark refractor, once the largest telescope in the world. The well-known U.S. Geological Survey geologist Grove Karl Gilbert, fresh off his famous—or infamous—expedition to Coon Butte (now known as Meteor Crater), in which he concluded that this well-preserved impact crater was actually the result of a volcanic steam explosion, used this telescope in 1892 to make "field" observations of the lunar landscapes. This led to his seminal (but not widely appreciated) work about the origin of lunar craters. At the time, most

---

1. Peter H. Schultz, "Shooting the Moon: Understanding Lunar Impact Theories," *Earth Sciences History* 17 (1998): 92.

astronomers still believed that they were volcanic, though Gilbert—rather ironically, given his conclusions about Coon Butte—produced compelling arguments in favor of an impact origin. Many years later, I received the G. K. Gilbert Award from the Geological Society of America and would recall being at the telescope "with Gilbert."

Even though I majored in astronomy, I had to choose what I was going to do with the rest of my life: astronomy, geology, art, or medicine (my father's choice). I decided to choose my passion and the most challenging path. I applied for graduate school in astronomy. Near the end of my senior year, however, the prospect of going to Vietnam became very real. Seniors were being bused to a school in a nearby town, where they would undergo their preliminary physical (if you were around then, you know what I mean). I worried that I was short enough to be chosen as a tunnel rat (someone who jumped into enemy tunnels) but learned that someone who knew radiative transfer was thought to be an asset to the country. As long as I was in good standing in graduate school, it shouldn't be a problem. The Astronomy Department was sensitive to this issue and supportive. Nevertheless, at least two of my graduate student friends committed suicide, in part because of this stress. When the Draft Board started eliminating certain deferments, my documented history of severe and chronic asthma reclassified my deferment to 4F. Who knew my asthma would become an asset?

For graduate school, I went to the University of Texas at Austin. I was interested in observational cosmology and initially worked with Gérard de Vaucouleurs on barred spiral galaxies as a possible strategy to improve the Hubble constant. A fellow student also working with him was given a C in one of his courses and told he wouldn't make it in graduate school. So, he moved to the Engineering Department and finished his PhD in celestial mechanics three years later. From this experience I learned the importance of finding the right combination of discipline, aptitude, passion, and unique skill sets. I also appreciated the fact that some professors are better to know than to work under, and that recognizing the distinction can make all the difference between success and failure in one's career.

My first year was spent preparing for qualifying exams and analyzing radio-astronomy data. But in the fall of 1966, the Geology Department hosted a week-long workshop about the Moon. Amazingly, the workshop's speakers included impact specialist Nick Short and volcanologist Jack Green. Nick (from NASA Goddard in Greenbelt, Maryland) specialized in shock metamorphism. I knew about Jack because he wrote an excellent history of lunar studies, which I referenced in my senior thesis. He was also well known for proclaiming that the only impact craters on the Moon were made by the Ranger 6, 7, 8, and 9. When Jack walked down the airplane ramp in Austin, he reportedly held up a new oblique view of the crater Copernicus from Lunar Orbiter 2 and proclaimed that he was right—it's a volcanic caldera! My senior thesis had come alive.

Only a month later, my future career in astronomy would be unexpectedly detoured. While on a flight to Austin, I was reading an article in *Sky & Telescope* about the planned Apollo landing, still two years away. A gentleman sitting next to me asked to see the article, and I obliged. He then mentioned that he was in the UT Geology Department and invited me to visit his office, which of course I did. This was J. Hoover Mackin, a legendary teacher and geomorphologist (whom some considered a maverick) with several classic papers still cited today. He was also training astronauts for the upcoming Apollo landings. After a long sit-down, he asked if I would like to look at the new photographs of the Moon from the Lunar Orbiter mission. Of course, I said yes; so, he gave me a thirty-four-kilogram box of twenty-by-twenty-four-inch photographs, which I carried about three-quarters of a mile to the car. After working on my assignments for astrophysics each evening, I started pouring across each photograph with a magnifying glass. When I returned to his office with the box, he gave me another, and then another. Later he asked if he could look at the Moon through the small telescope on top of the physics building, where the Astronomy Department had an upper floor. Mackin was well known as an energetic geologist who would shame his students with his fast pace in the field, but I noticed he was out of breath as we climbed the five flights of stairs. I didn't know that he had just had a cardiac pacemaker implanted. (At the time, pacemakers were rather novel; the first ones had been introduced only seven years earlier and were not as dependable as they are today.) He just couldn't slow down, even after surgery.

The next year (1967), Mackin asked if I would be interested in working on an atlas of lunar features for NASA. This would involve identifying what features were in which photograph. As I began setting up the matrix of photos and features, I realized that this was a bit silly: there's a crater everywhere, and I was more interested in the processes shaping the surface. Although I did the classification, I also took notes about what could have formed these features. This led to more geology courses while taking astrophysics in astronomy. In these courses, I met an undergraduate, Ruth Fruland, who was also interested in the upcoming lunar missions, and Suzanne Weedman, who was the wife of a new faculty hire in astronomy. I'll return to them later.

During the first year, I took my first trip to McDonald Observatory, and three memories still stand out. First, Harlan Smith took me outside on the observatory catwalk around the dome, pointed out Omega Centauri, and talked about the geology of the surrounding terrain (his son would become a geologist). Second, the McDonald 82-inch reflector was the largest telescope I had ever seen. Sparking relays would light up the windows every time the dome was repositioned for viewing through the slit. From outside at night, this seemed like a scene from a horror movie: something onerous was happening inside this giant round-top mansion. And third, I saw my first tarantula (under a rock), a very large black widow (just inside the door of the 36-inch telescope dome), and a scorpion (ran across my path one afternoon). I actually trapped all three and set up bets on who would win in a tag team match (guess who won). Although my concern about the draft forced me to stop spending time painting and drawing, I couldn't resist capturing the telescopes on Mt. Locke: the legendary 82-inch McDonald telescope and the 107-inch telescope (now the Harlan J. Smith telescope), which was under construction at the time (figure 4.6).

Much later I would realize that my art background enhanced my research in various ways. First, it created a strong visual memory, which would prove important for recognizing and remembering planetary surface features. Second, my approach to painting required complete immersion—not just dabbling. This applies to research as well. Third, studying art would help me design impact experiments. And fourth, it helped me visualize experimental outcomes, whether in the laboratory or during NASA's Deep Impact and LCROSS missions.

During Thanksgiving of my first year in grad school, I proposed to my high-school sweetheart (and meteor spotter) under a live oak tree near campus. We were married as soon as she completed her degree that spring. Back then, we lived on $119 a month as a research or teaching assistant, with $90 going to rent. (That sounds impossible, but gas was below twenty-five cents per gallon, and a good steak cost less than fifty cents a pound.) When we returned to the UT-Austin campus fifty years later, I surprised her by returning to that very tree.

FIGURE 4.6 Watercolors showing, *left*, the McDonald 82-inch reflector, which when first built (1933–39) was the second largest telescope in the world, after the 100-inch reflector at Mt. Wilson; and, *right*, the 107-inch telescope (now called the Harlan J. Smith telescope), which was under construction 1966–68. These watercolors were made by the author while not observing in 1967. Courtesy of Peter H. Schultz.

I continued working on the moon atlas in 1968, while deeply involved in classes and passing the dreaded qualifying exams, which involved two days of written tests and an oral examination. After passing, I was on my way. But, while rushing to the hospital in Houston on August 12, 1968, Mackin passed away from a staphylococcus infection during surgery for a replacement pacemaker. I sat on the back porch of our little rented house in Austin and just stared. It just didn't seem possible, especially on the same day as the maximum of the Perseid meteor shower. Ever since, the Perseids remind us to toast to Dr. Mackin and his life.

The passing of Mackin could have meant the end of the lunar orbiter atlas project, except for another UT-Austin geology professor, Bill Muehlberger. Bill took on not only Mackin's grants (including the project I was working on) but also his efforts on the Apollo landing, now less than a year away. He allowed me to continue working on the project in Mackin's large office, with a long wall of books, pamphlet files, and piles of papers. I would later adapt his filing system, that is, the stratigraphic method involving layers of paper. When they finally cleaned out Mackin's office, Bill set up a separate room across the hall.

Bill was also an impressive geologist and individual. During this time, I was clearly doing much more than just making an atlas. Bill not only encouraged this approach but had worked out an arrangement with NASA to publish the final product as a book through University of Texas Press. He shielded me from the hectic activities involved with the Apollo landings. This may seem cruel to graduate students often deeply engaged in NASA missions now, but it was a blessing for me. The first three Apollo launches were separated by only four months. Between each mission, the geology team had to publish results from the prior mission and plan for the next, not to mention advocating for their favorite place at site-selection meetings. Although launches were separated by about eight months after Apollo 13, this pace was still not conducive for working on a PhD thesis. Nevertheless, Bill invited me to sit in on planning meetings for future landing sites. At dinner, my wife and I were lucky enough to sit with the Apollo 13 astronauts (Jim Lovell, Fred Haise, and Jack Swigert). This was only a month after they had returned, and I had to ask what they saw as they passed over the lunar farside. They responded that they were "busy." I suppose they were, rather!

Harlan Smith suggested that I attend the Gordon Research Conference in Tilton, New Hampshire, in the summer of 1968, my first professional conference. The people and personalities made a lasting impression, not to mention the science. Lunar orbiter photographs had just become available, but it was six months from Apollo 8 and a year from the first landing. Here I met several legendary figures in early space exploration: Al Cameron (cosmogonist), Mike Duke (NASA, Johnson Space Center), Hal Masursky (USGS), Mike Dence (Ottawa Observatory), John Wood (Harvard), Bill Quaide (NASA Ames), and Don Gault (also NASA Ames). Don would later become my mentor at NASA Ames during my postdoctoral appointment, as well as a twenty-five-year partner in cratering crimes using the Ames Vertical Gun Range (AVGR), a facility he designed in the midsixties. In addition, I met newcomers, including graduate student Clark Chapman (a contributor to this book), who was then at MIT, and Jim Head at Bell Labs. Clark had a car and offered to drive me to a nearby town to look for antique cameras, which my wife and I were now collecting (this trip actually led to finding a daguerreotype camera several years later). After graduating, I almost joined Clark in Tucson. Jim went from Bell Labs to the Houston Lunar Science Institute, where he briefly served as interim director, before joining the faculty in the Department of Geological Sciences at Brown University—and where, as I little dreamed then, I would also end up. The Gordon Conference proved to be a seminal event in my career; I enjoyed not only the open discussions about science but also the interactions and informal banter in a relaxed atmosphere. (I hasten to add that this was be-fore the competitions for research grants and battles between disciplines became quite so intense.)

I watched the Apollo 11 landing in 1969 with new assistant professor Dan Weedman and wife, Suzanne (one of the UT-Austin geology undergraduates mentioned previously). I set up my large reel-to-reel tape recorder and cam-era in front of their TV set. I recorded every station's live coverage, before and after the landing. On the twenty-fifth anniversary of the first Apollo landing, we reunited in Washington, D.C. Dan was now NASA's discipline scientist for astrophysics while Suzanne was working at USGS Langley. This happened to coincide with comet Shoemaker-Levy's collision with Jupiter. Ironically, we went to an open house at the Naval Observatory and observed the comet corpses lying out on Jupiter's upper cloud decks through the ven-erable old 26-inch refractor.

The late sixties was the beginning of meetings about lunar and planetary exploration. While I was at the Naval Observatory, staff scientists brought along a few interns to Cornell for the American Astronomical Society (AAS) meeting, where the first results from the Surveyor 1 mission were presented. This was the first of many "first" conferences focusing on the Moon and plan-ets: the first Division of Planetary Sciences of the AAS (Austin); first Plane-tary Geology Program Principal Investigator's meeting (Flagstaff); first Lunar Science Conference (Houston); first international conference on Mercury (Houston); first international conference on Mars (Pasadena); first conference on impact cratering (Flagstaff); and later, the first conference seriously ad-dressing the role of an impact on the great extinction about sixty-six million years ago.

Of all these, the first Lunar Science Conference stands out in many ways. This meeting was a program review for funded researchers, not graduate students like me. But I was able to sneak in with the equivalent of a Willy Wonka golden ticket through someone in the Public Affairs Office at NASA's Manned Spacecraft Center (MSC, later renamed the Johnson Space Cen-ter, or JSC). This someone was the mother of that other UT-Austin geology undergraduate mentioned above, Ruth Fruland, who was working at MSC as a NASA Co-op. To save me money, Muehlberger asked if I wanted to room with him. Filled with anticipation, I walked to the convention center from the hotel under one of those crisp January days with a deep blue sky in downtown Houston. I had my camera with me but thought it would be unprofessional if I took a bunch of photos, especially since I wasn't sure if I should be there anyway. Besides, I had loaded the camera with color film that had a sensitivity requiring timed exposures in the meeting room. So, I took one photograph of the convention center from the outside and another of the registration desk inside—that's all.

In the summer of 1969, another geology professor, known by his students as Dr. I (the well-known geochemist F. Earl Ingerson), asked if I wanted to look at recently released images from the Mariner 6 and 7 mission to Mars. Of course the answer was yes. In 1965, Mariner 4 had returned images showing only a few craters with nothing else very interesting. The first published reports from Mariner 6 and 7 images supported this conclusion and suggested that Mars was just like the impact-battered Moon. Some researchers mapped many linear features and proposed a global fracture system, something that also had been proposed for the Moon at that time. With cross-section methods, I was able to show, however, that most of these "fractures" were actually products of coherent noise introduced while processing the image. Accounting for such artifacts, it was clear to me that Mars had more than just craters: it also had narrow sinuous valleys, chaotic terrains, and what appeared to be outflows from giant canyons. I told this to a newly arrived professor at UT, Vic Baker, while on a plane ride in 1971 and asked if the outflows could be analogous to the Washington Scablands, a process that had fascinated me since introductory geology. This was before Mariner 9, when such features would become obvious. Vic would later become rather famous for championing this hypothesis. I doubt he recalls this encounter, but it remains vivid in my mind because I had lost the chance to be the first to propose this idea. Unfortunately, I had to finish my thesis before publishing what I found.

In 1969, Ingerson asked if I would give a talk about my results at a Principal Investigator's meeting in Flagstaff, Arizona (my first professional talk). This meeting was the precursor to the Planetary Geology and Geophysics Principal Investigator's conferences, which would parallel the Lunar Science Conference in Houston for more than two decades. After giving my talk, Don Wise (U. Mass) approached me and said, "You were just waiting to give this." I didn't quite understand what he meant, but apparently I had unwittingly just trashed work by other investigators in the audience.

By late 1969, I realized that I had better find a PhD topic. So, I started working on the midinfrared Christiansen bands applied to spectra of Mars with the late Bruce Ulrich. Instead, my joint advisers (Smith in astronomy and Muehlberger in geology) proposed that I submit my lunar project as my thesis. Even though Smith tried to convince me that "perfection is the enemy of the good," I had to go back and add more detail and structure.

My questions were all process based, and I formulated various scenarios (alternative working hypotheses), an approach that naturally evolved from my senior thesis at Carleton. In addition to detailed descriptions, the thesis involved discussing selected features, making Polaroid photographs, grouping and photographing them on a page, and finally developing and printing each page in the darkroom. By the time I finished, my thesis had grown to two thick volumes, in which each page of photographs of classified lunar features was accompanied by interpretations with summary sections. Before 1972, a submitted thesis required original typed pages on archival paper; photocopies were not allowed. In addition, all photographs had to be originals, including all the versions for my five committee members. Fortunately, the university had just changed their requirement for original typed pages but still required original photographs. This meant that I had to produce over five hundred pages of photographs for each copy given to my committee members, in addition to the submission to the university. In other words, I photographed, processed, and printed more than 3,500 photographs in the darkroom, in addition to all the Polaroids. If nothing else, I could use these two thick bound volumes to help me see over the steering wheel in our big, used 1967 Buick LeSabre.

After my defense, I thought I was all done. In a month, however, I noticed that some of the photographic pages were beginning to turn yellow. I knew what that meant: someone had not thoroughly washed their photographs, which left fixer stains in the cloth belt wrapped around the print-drying drum. The university wouldn't accept a stained, nonarchival thesis; all the prints would have to be redone or cleaned. Over the next three months, my wife would come directly from work, and together at night (5 p.m. to 11 p.m.), we would send each photographed page back through the stop bath, fixer, and wash (with a clean belt, of course). I finally gave each committee member their copy of the final thesis with a signed front page of the first volume. Years later, the second volume of my thesis showed up for sale. I wanted to believe that one of my committee members chose to hold on to the first signed volume (more likely he didn't want to be outed).

After submitting my "clean" thesis, I received a National Research Council postdoctoral fellowship at NASA Ames, working with Ron Greeley and Don Gault. I wanted to go to Ames because I felt that my observations of small surface features on the Moon contradicted current models of impact degradation. They allowed me to reanalyze their published (and unpublished) crater statistics of various locations on the Moon. By combining my detailed knowledge about the Moon and visualizing patterns in the statistics, I was able to rapidly publish several papers based on my thesis: one about the effect of regolith thickness on crater statistics; another about the implications of small features on the maria that should not have survived three billion years; and the effects of impact-induced seismic shaking. I also expanded other observations in the thesis about floor-fractured craters and the role of antipodal shock effects from large basins on the Moon and Mercury. Finally, after another two years, my thesis was published as a book, *Moon Morphology: Interpretations Based on Lunar Orbiter Photography*.[2]

As I reflect, I realize how much my diverse interests (astronomy, geology, art) influenced my topics of interest and approach. For example, I discovered that I compose a manuscript just like an oil painting: first sketching my thoughts, rearranging, removing layers, rediscovering, before finally reconstructing the final work. Although this approach is inefficient and time consuming, it always leads to new insights and ideas. I also realized that the smallest of features (or most obscure) can have the biggest implications. Finally, I learned to appreciate (and respect) the differences in approach between different disciplines and the value of having two places to work (geology and astronomy) with very different cultures: one drank wine and martinis, the other beer—you can figure out who drank what.

I've been lucky because over the last fifty years, I've been able to ask my own questions and look for the answers. When I started, the planets, asteroids, and comets were just astronomical objects; now they belong to the geologists. I've now been (so to speak) to the Moon, Mercury, Venus, Mars, the asteroids, comets, and the satellites of the outer planets. On Earth, I've

2. Peter H. Schultz, *Moon Morphology: Interpretations Based on Lunar Orbiter Photography* (Austin: University of Texas Press, 1976).

worked on eight different impact sites in Argentina and investigated a crater in Peru in 2007, as well as glasses generated by a low-altitude air blast
over the Atacama Desert. I've gone into sinkholes formed above the impact
crater on the Yucatán from the asteroid that killed the dinosaurs. And I was
part of a mission that slammed into a comet in 2005 (Deep Impact) and
another that ejected water ice from beneath the cold shadows at the lunar
poles (LCROSS). While I am most certainly from "the space age generation,"
my fascination began well before.

It started with my photograph of a comet in 1957 and would lead to my
trekking across the Atacama Desert. These two life events actually connect.
Small meteoritic fragments found in the Atacama glasses closely resemble
minerals in samples returned from NASA's Stardust mission to a comet. In
a way, my earliest amateur life and professional life merged. And in 2022,
I returned to the Prairie Astronomy Club in Nebraska to give a talk about
defending the Earth from events like the one that generated the glasses on
the Atacama. This was the same night that NASA's DART mission slammed
into an asteroid (six of my past students made significant contributions to
this mission). But there's more. Remember when my high-school girlfriend
and I watched the Perseid meteor shower on a farm south of Lincoln? The
farmer's wife (now ninety-three) surprised us by coming to the talk as well.
We all watched together in real time. A special night.

5

From Pre-Sputnik to a Career in Planetary Science

CLARK R. CHAPMAN

As the Sun moved westward in the sky, I shifted a yellow crayon-colored paper disk right to left along a string in my bedroom. That's how my mother described one of my activities as a three-year-old. One evening, as my father planted a lawn behind our new house in a Buffalo suburb, I looked up and was transfixed by constellations in the darkening sky. Perhaps a year later, I convinced a couple of neighborhood kids—Mickey and Ann Newton—to form the NewChap Astronomy Club, for which I made a hand-drawn "book" about the planets and stars, using shiny reward stick-on stars to form constellations.

My father was a physicist, specializing in atmospheric electricity. Occasionally he launched weather balloons from our front yard. He also constructed a 2-inch telescope on a tripod, using lenses in a cardboard tube. One evening he pointed it at Saturn and its rings, which I gazed at in awe. Many years later I would be awed again admiring the subtle hues of the ringed planet through the famed 60-inch telescope on Mt. Wilson, north of Pasadena, during a still night of perfect seeing. As a kid, I was fascinated by Chesley Bonestell's space art in *Collier's* magazine, and I also read Robert Heinlein's *Red Planet,* a science fiction book featuring students and an intelligent furry ball. Science fiction got me enthused about astronauts, so I've always supported sending human beings into space and never agreed with some colleagues that only robots, not people, should explore asteroids and planets.

In June 1954, my dad took me on a trip to a mosquito-infested field near Minneapolis for us to watch the total solar eclipse while he made electrical measurements and took a series of photos of the waning and waxing phases of the Sun as the Moon crossed over. Flying back to Buffalo, we stopped for a day in Chicago and enjoyed an Adler Planetarium show, a superior venue compared with the small domed ceiling at the Buffalo Museum of Science.

I was in my bedroom the afternoon of October 4, 1957, when my radio reported that the Soviet Union had launched the satellite Sputnik. My dad immediately began calculating how to see it. Sputnik itself was too faint to see, but its booster rocket remained in orbit as a bright moving star. The two local non-Polish-language daily Buffalo newspapers contacted my dad, who remained their go-to expert on things in space for years thereafter. My dad also provided Sputnik viewing information for me to print in my weekly neighborhood newspaper, the *Kingsgate Tribune* (circulation about one hundred for a two-cent hectographed sheet). My October 13 edition reports that I and several others had seen the Sputnik booster early that morning.

My dad wanted to study the shapes of the Russian orbiting rockets, and so, in 1958, he purchased a 10-inch Cave Astrola reflector, a telescope I still have in my barn in Colorado (figure 5.1). (Some years later, I found myself working next to Alika Herring, who ground the mirror for "my" telescope.) The telescope wasn't useful for catching rapidly moving satellites, but I adopted it to try to observe Mars during the red planet's autumn opposition. Autumns are always cloudy in Buffalo, but I would track holes in the clouds and try to spot the planet as a hole passed by. Thus began several years of my dedicated middle-of-the-night backyard observing, chiefly of Mars, Jupiter, and lunar craters.

In 1959, my mother developed a rare tumor in her eye, which American doctors could do nothing about, but she had an opportunity to receive experimental radiation treatment in London. (It was successful—she passed away less than a decade ago at age 101.) So I was dragged off to England for several months, away from my telescope. There, I borrowed a library copy of Ray Bradbury's *The Silver Locusts* (slightly different from the U.S. version, *The Martian Chronicles*). I didn't like Bradbury's description of "the blue sands of Mars." Years later, when I chanced to chat for a while with Bradbury, I hesitated to confront him about planetary hues, which have always been a sore point for me. In third grade I was mistakenly taught (though maybe I misinterpreted) that Saturn was yellow because it was covered with sand. Even

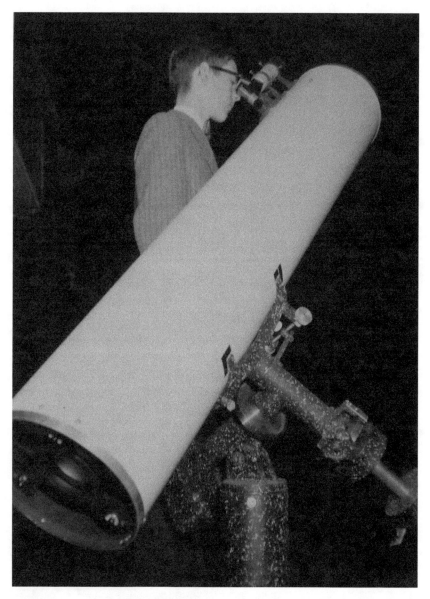

FIGURE 5.1 Teenage Clark Chapman looking through the eyepiece of his 10-inch Astrola telescope. Courtesy of Clark R. Chapman.

as I entered college, I retained the cognitive dissonance between the yellow sand and my objective knowledge that Saturn is enveloped by a cloudy, hazy atmosphere. And in my 1999 Sagan Medal talk in Italy, I criticized the vertically exaggerated and golden colorized Magellan radar images of landforms on Venus; some listeners didn't like my complaint. Quite recently, a bluish Martian sand dune was imaged by Mars Reconnaissance Orbiter, partly vindicating Bradbury's famous imagination.

Back in Buffalo, I read more science fiction. And I borrowed amateur astronomy books from my local library, such as *Guide to the Moon* by Patrick Moore. (I met Moore decades later as he reported the Voyager Saturn encounters for the BBC from the Jet Propulsion Laboratory, or JPL). I soon joined the British Astronomical Association of amateur astronomers. I also discovered the American amateur Association of Lunar and Planetary Observers (ALPO) and began subscribing to its journal, the *Strolling Astronomer*, edited by Walter Haas. Early in 1960 I learned, through *Sky & Telescope* (*S&T*), of a National Science Foundation–sponsored summer program in astronomy and rocketry. Haas and Professor Paul Engle of Pan American College taught the thirty-two boys and girls from around the country during the six-week program.

Flying to Edinburg (near McAllen by the Rio Grande in extreme south Texas) was my first solo trip away from home. Students were housed in the Echo Motor Hotel and took a bus to and from the college. After classes, we swam in the pool, listening to blaring music of the time, such as Roy Orbison's "Only the Lonely." I played duets on the piano in the hotel's lobby with fellow student Tom Constanten. Tom later became an early keyboardist with the Grateful Dead and is surely the only star in Cleveland's Rock and Roll Hall of Fame whose published autobiography includes drawings of lunar craters. During evenings, many of us used Engle's Moonwatch station or observed with his 17-inch telescope. I marveled at Omega Centauri, the famous star cluster visible only from the southern extremities of the United States. One morning the class launched a small rocket and spent much of the day scouring the flat plains (the highest topography consisted of termite mounds) to recover it.

The high point of the summer was a camping trip to the top of 10,300-foot Cerro Infiernillo, 250 miles away (south of Monterrey), to scout it as a possible observatory site. The lustrous Milky Way was spectacular, with car lights far below along the Pan-American Highway. Many students were sick from bad food, but another guy and I were well enough to chop down one pine tree, as a start for the would-be observatory. It was an auspicious group of kids, some of whom I kept up with over the years. Tom Stoeckley actually moved to the small Colorado town (population ~2,000) where I live. Bob Havlen became the executive director of the Astronomical Society of the Pacific in the nineties. I met with Paul Knauth several times while he was professor of geochemistry at Arizona State University. Tom Constanten still comes through Colorado occasionally; my wife and I met with him a while back while he was touring with Jefferson Starship. And for Christmas 2021, he sent us a new CD of his music.

Following my Edinburg adventure, I became very active in the Association of Lunar and Planetary Observers. In September 1960, I attended its convention in Haverford, Pennsylvania, where I met Bill Hartmann (see his chapter in this book), who had an influential role in my career a few years later. I traveled to ALPO conventions in Detroit, Montreal, and elsewhere, and I spent countless nights mapping Jupiter and Mars, mailing my drawings to Haas to print in the *Strolling Astronomer* (figure 5.2). In high school, I persuaded my health science teacher that it would be "healthy" for some of us kids to

FIGURE 5.2 Pages from Clark Chapman's observing logs in 1961 and 1962, showing his drawings of Saturn, Jupiter, and Mars. Courtesy of Clark R. Chapman.

leave our classes and watch the Project Mercury launches on the student union TV. My 1962 science project—a map of Mars and studies of Jupiter's belts and zones—earned my selection as one of forty national winners of the Westinghouse Science Talent Search (STS). While in Washington, D.C., we gathered in the Oval Office, where I briefly chatted with President Kennedy. Recognition by the STS no doubt helped me get accepted for college by Harvard, the University of Michigan, and Cornell; for both better and worse, as it turned out, I chose Harvard.

I searched for a summer job before the fall 1962 freshman semester. Bill Hartmann helped me apply to Gerard Kuiper's new Lunar and Planetary Laboratory (LPL) at the University of Arizona, where Hartmann was a first-year graduate student. Kuiper was the dominant planetary astronomer of the time, and he hired me to join his group creating a map and catalog of lunar craters based on his recently published *Photographic Lunar Atlas*.[1] I flew out to Tucson in June and stayed in the main downtown hotel. The next morning, I put on a suit for my first-ever job and walked over a mile to the university's Steward Observatory (where else would a lunar lab be?). There I was told that the LPL was actually way back in the Institute of Atmospheric Physics building I had passed (why LPL was not part of astronomy on campus is discussed elsewhere in this book). I was sweating by the time I climbed up to Kuiper's offices, where I met my new boss, Dai Arthur, and other lunar specialists, like Chuck Wood (another contributor to this book), Ewen Whitaker, and Alika K. Herring (figure 5.3). I quickly realized that nobody wore suits in the desert Southwest. In fact, that day was the hottest day ever recorded until then in Tucson (110°F).

The crater catalog, finished a few years later, remained the definitive catalog of frontside lunar craters until just the past decade. During my final summer in Tucson in 1964, the Ranger 7 lunar probe was launched; it took the first high-resolution pictures of tiny craters as it crashed into the Moon. Kuiper and Eugene Shoemaker were experimenters on the mission, and I joined an attempt to observe the probe's lunar crash using telescopes on Kitt Peak. I met other well-known planetary astronomers and geologists during my three summers in Tucson, including Audouin Dollfus, who gave a talk at a blackboard and, memorably, picked up a piece of chalk, snapped

---

1. Gerard P. Kuiper and Ewen A. Whitaker, eds., *Photographic Lunar Atlas* (Chicago: University of Chicago Press, 1960).

FIGURE 5.3 Alika K. Herring at work in the Lunar Lab, early sixties, drawing a lunar limb region. Earlier he had sketched pictures of Mars during its fifties oppositions. Meanwhile he configured telescope mirrors, including mine, at Cave Optical. A native Hawaiian and steel drum player, Herring was the first to site-test mountains in Maui and the Big Island for observatories that would blossom into some of the finest in the world in later decades. Courtesy of Clark R. Chapman.

it in two, and—holding a piece in each hand—started at the top and drew a symmetrical circle coming back together at the bottom.

Entering Harvard College in 1962, I soon realized it didn't match my desires. I wanted to major in astronomy but was told that meant taking courses in physics and math, not astronomy. Taking Wendell H. Furry's thermodynamics brought me none of the joy I felt from observing Jupiter. I had one small nonscience class taught by a grad student named Erich Segal, who would go on to write the novel and screenplay for the blockbuster movie *Love Story*. But whether I worked on a paper for weeks or dashed something off in the half hour before class, Segal never graded me higher than a B– nor lower than a C+.

I took a couple of journalism classes, which were more interesting. Since making my youthful debut as a journalist with the *Kingsgate Tribune*, I had worked on school newspapers and had continued to add to my newspaper collection, often by trading a copy of my newssheet for a copy of their paper. A Columbus, Georgia, editor repeatedly encouraged me to become a journalist. Given my dislike of Harvard's approach to astronomy, after my second summer in Tucson, I took a leave from Harvard to see whether physics and math classes at the University of Arizona might be better, or whether I should go into journalism instead of astronomy (figure 5.4). Arizona accepted only a few of my Harvard credits, so I entered U of A as an "upper freshman" in autumn 1963, while continuing to work part time on the Lunar Lab's crater catalog. I was measuring craters when someone came in and told us that President Kennedy had been shot.

As it turned out, I enjoyed and did well in my physics and math classes, so I was persuaded to return to Harvard for the 1964 spring semester as an astronomy major. Instead of straight physics and math, however, I enrolled in various classes related to the planets, like geology, geophysics, and physics of atmospheres and oceans. These were interesting, but as I eventually realized, the Harvard geologists were behind the times; in one class, the professor "disproved" continental drift, though he allowed a graduate student to speak to the class in its favor. To amuse myself, I was granted access to the famous old Alvan Clark 15-inch Great Refractor at Harvard College Observatory, and when time permitted, I continued my planetary observing. But at the end of one night, I was closing the dome's slit shutter when a chain snapped, and it crashed down; my permission to use the telescope was revoked. I did enjoy writing a term paper on globular star clusters, overseen

FIGURE 5.4 A montage of the author's interests in journalism. His earliest "newspaper" was printed just for his parents in 1953. Small pictures show him and his parents at his typewriter and "printing press" during the time he produced the *Kingsgate Tribune*. Note that one headline reports his observation of the Sputnik booster. He also wrote an article for a Tucson alternative newspaper about the NASA planetary funding crisis in the early eighties. Still later, he edited a homeowner's association newsletter. Courtesy of Clark R. Chapman.

by Cecilia Payne-Gaposchkin. She was a famous female astronomer and a strong personality. During one of my sessions in her office, her door cracked open, and her husband, Sergei, poked his head in. She shooed him away like a bothersome fly.

A silver lining to my Harvard experience was Carl Sagan's course The Planets. I got only a C+ in his class; I missed the final exam because a Harvard Health Services doctor prescribed me medicine that did the opposite of what I needed, so I was bedridden that day. Despite the grade, Sagan appreciated my studies of craters. In summer 1965 I was hired by Gerald Hawkins to work at the Smithsonian Astrophysical Observatory (located at Harvard Observatory) to analyze Hawkins's *Boston University Catalog of Lunar Craters*. I learned FORTRAN programming but was stunned one day when my overnight computer run resulted in a cart loaded with a two-foot-high stack of output, the same line of text printed on each page because of my faulty do-loop. My office was just a few doors from Sagan's, and he invited me to join him and grad student Jim Pollack in analyzing the first pictures of Martian craters from Mariner 4, which resulted in a paper by the three of us published in the *Astronomical Journal*.[2] I continued to interact with Sagan over the years until his premature death. One evening in 1974 in Pasadena, I had a Japanese meal with Carl, during which he asked me (as he probably queried many others) whether he should expand his role in popularizing astronomy. I told him it would be a good choice. After dinner the two of us went to see a movie, *The Four Musketeers*. Carl Sagan's class, and later interactions with him and Pollack, clinched my decision to pursue planetary science, not journalism, as a career.

While working in Tucson, I read an article by Raymond Hide of the Massachusetts Institute of Technology (MIT) about the nature of Jupiter's Red Spot. Having often observed Jupiter, I wrote to Hide. He later hired me to work for him during the summers of 1966 and 1967 in Las Cruces, where New Mexico State University (NMSU) was taking and cataloging Jupiter images as part of its participation in the International Planetary Patrol Program. By this time, Walter Haas lived in Las Cruces, as did famed astronomer Clyde Tombaugh, the discoverer of Pluto; indeed, Clyde's office was next to mine. His main research interest just then was Mars. I also got to know the leader of the NMSU research group, Brad Smith, who seemed, and was, the reserved, proper Bostonian. He later evolved into the gregarious leader of the imaging team for the Voyager missions to the outer planets.

Throughout the sixties, I worked with *S&T* editor Leif Robinson, Dale Cruikshank (another chapter author in this book), and others to create the *ALPO Observing Manual*, with chapters by many amateur experts on the individual planets, comets, and techniques of observing, as well as a chapter on the theory of visual observing by Charles Giffen. We finished the book but couldn't get any prospective publisher to close the deal. One day in the seventies, I walked into a small shop in Tucson and actually found a copy of our book for sale. It was simply a reproduction of the pages of our typed text. The book was actually listed on Amazon.com in the late nineties but not available. Unfortunately, all that effort never really helped the amateurs for whom it was intended.

---

2. C. R. Chapman, J. B. Pollack, and C. Sagan, "An Analysis of the Mariner-4 Cratering Statistics," *Astronomical Journal* 74 (1969): 1039.

Harvard and I still didn't get along well. I did a project reconciling a half century of Mercury drawings, gleaned from the observatory library but also my own observations, with the then new rotation period for the planet determined by radar (and I wrote an associated article for *S&T* with Dale Cruikshank); but the Harvard Astronomy Department wasn't impressed. The esteemed comet scientist Fred Whipple was on a committee that quickly denied me honors (Whipple was very friendly toward me some years later, having forgotten me as a Harvard undergraduate). The Harvard Advisory Office for Graduate and Career Plans concluded that I couldn't get into any graduate school and advised me not even to apply—so after graduating in 1967, I applied to nowhere. That summer, the phone rang, and it was Hide, who told me that he had arranged for me to meet with Henry Houghton, then head of the MIT Department of Meteorology. Hide, who was in the Department of Geology and Geophysics in the same Building 54, wouldn't tell me what my appointment was about. I met with Houghton, who immediately invited me to become a graduate student in the Meteorology Department. I signed my name on a sheet of paper and was admitted to MIT without even applying!

That autumn I had an office, as a graduate assistant researcher, in an upper floor of Building 54, where I continued researching Jupiter's cloud belts while taking meteorology classes leading to my master's degree. I began attending professional scientific conferences. In April 1968, I climbed into a van with other MIT students, and we were driven to Washington, D.C., for the annual spring meeting of the American Geophysical Union. It was the day after Martin Luther King Jr. had been killed, and rioting was under way. We drove by Wilmington, Delaware, and watched smoke rising above the city. Soldiers turned us away as we tried to enter D.C., and we were forced to spend the night in Virginia. We managed to argue our way into the city the next day. These memories returned during the siege of the Capitol in early January 2021.

By happenstance, I learned of another planetary meeting later in 1968 in Austin, Texas, and I went to it to talk about my master's thesis—which was really the culmination of my Jupiter studies that began back in high school. That meeting became the pre-inaugural meeting of what the next year would become the new Division for Planetary Sciences (DPS) of the American Astronomical Society. The DPS is still the main professional society of planetary scientists, and I'm the only person who has attended every single DPS meeting, thanks to good luck in never being too ill to attend. I was actually elected as chair of the DPS in the early eighties during a dark period for planetary science. Newly elected President Ronald Reagan's director of the Office of Management and Budget, David Stockman, wanted to zero-out NASA's budget for planetary exploration over three years. I worked on Capitol Hill with Carl Sagan and others to overturn that decision.

By 1968 the Vietnam War was building up, and I wanted nothing to do with it. My professors suggested that I become a full-time researcher and get an occupational deferment from the draft, which I did. Since the air force sponsored my Jupiter research (why?), I rationalized that I was doing my peaceful part for the military with my studies. Anyway, looking toward my future, I applied for PhD programs in both the downstairs MIT Department of Geology and Geophysics and to Caltech. I was accepted by both, and in spring 1968 I flew out to Pasadena to learn what Caltech might be like. Hunting for the right building, I encountered Professor Peter Goldreich jogging on the sidewalk, and he pointed me toward Mudd Hall. There I spoke with Bruce Murray and others but was advised that while I could major in planetary science, I had to do a minor in astrophysics, so I was directed upstairs to talk with Jesse Greenstein. He was so intimidating that I decided then and there that I'd go to MIT instead. Before I left Pasadena, I went to the Caltech student center to watch a TV speech by President Lyndon Johnson, in which he famously announced that he would not seek nor accept nomination for another term.

Compared with Harvard and the foreboding possibilities at Caltech, MIT was amazing. In my Meteorology Department office, I got a notice of a brown-bag lunch on a lower floor for anyone interested in the Moon. Gene Simmons had just returned from Houston and announced that he had a lot of funding for lunar research. He asked anyone who was interested to raise their hand, so I raised my hand. As a result, I got an entire room in Building 54, across from Simmons's office, filled with the huge collection of Lunar Orbiter photographs and funds to hire an undergraduate to help count craters. That remains my simplest proposal ever: I raised my hand. It was a great way to begin as an MIT PhD candidate in planetary science. (The Department of Geology and Geophysics soon changed its name to Earth and Planetary Sciences. Later still, it merged with the upstairs Department of Meteorology and is now the Department of Earth, Atmospheric, and Planetary Sciences.)

I had yet another opportunity for a planetary summer job in 1969. Irwin Shapiro was glad to hire me to work at MIT's Haystack Observatory, in an exurb of Boston, to analyze its radar echoes from the planet Venus. I was told that I needed a security clearance to work there. But I argued that no military secrets were associated with Venus. Shapiro accompanied me to meet with the vice president of MIT and supported my plea to be exempted from the requirement. The VP agreed, and I became the first person to work at Haystack without a security clearance.

As a new student, I was assigned to work with Tom McCord, who had just arrived as a new PhD from Caltech. He had built a special spectropho-tometer with a filter wheel covering wavelengths from the near-ultraviolet through the near-infrared. He suggested that I use it to observe the colors of asteroids to determine their mineralogy. That turned out to be my PhD thesis, although I wove into my research aspects of cratering of asteroid surfaces, given my long interest in craters. I had many trips, shipping the spectrophotometer to distant mountain observatories. My first observing run, in June 1970, was using the Hale telescope on Mt. Palomar, then the largest telescope in the world. It was downhill (at least in telescope aper-ture) thereafter. Later I took the spectrophotometer to Mt. Wilson, Kitt Peak, Mauna Kea, and Lowell Observatory, to measure asteroid spectra. McCord assembled a large group of planetary students and associated faculty under the name MIT Planetary Astronomy Laboratory (MITPAL). I formed many longtime associations with MITPAL colleagues. A few years ago I attended MITPAL's fiftieth anniversary in Cambridge.

One memorable episode concerned Saturn's rings. In late 1969 the *New York Times* reported Gerard Kuiper's finding that Saturn's rings were made of ammonia ice. A group of us at MITPAL, working with Hugh Kieffer of the University of California, Los Angeles (UCLA), compared the Lunar Lab spectrum of the rings with laboratory spectra of ices and realized that the rings were actually water ice. We wrote a simple draft article to that effect and sent it to colleagues at the Lunar Lab. Some weeks later, I was vaca-tioning in Tucson and learned that Kuiper had sent his assistants into the laboratory to measure ice spectra, realized that their rings spectrum was of water not ammonia ice, and had sent a note to that effect to *S&T* for pub-lication, but without mentioning the MIT/UCLA results. I arranged for a tendentious meeting with Kuiper in his living room that evening, while Mrs. Kuiper served me tea. Kuiper was miffed about MIT's invading his territory, telling me that he "owned" Saturn's rings. The next morning, a group of us (including Dale Cruikshank) gathered in Kuiper's office for a telecon with editors of *Sky & Telescope*, while one of our group (Carl Pilcher) showed up at *S&T*'s front door, a couple of miles from MIT. We negotiated changes to

S&T's story, properly reflecting our identification of water ice. The formal publication of our paper, by Pilcher et al., was rushed into print in *Science* magazine, facilitated by Frank Press, the head of our MIT department, and *Science* editor Philip Abelson.

Professors on my PhD committee at MIT encouraged me to finish up. After all, I had been around Building 54 too long, they said, perhaps unaware that much of the time, I was just a paid researcher, not a student. (Caltech would have had a different attitude!) The department experimented, with me and a few others, to give a colloquium about my thesis rather than take an exam. Shapiro, away on sabbatical when this experiment was initiated, didn't approve, but he attended my "thesis defense" nevertheless. Partway through the Q&A after my talk, a baseball crashed through and shattered an auditorium window, bringing my thesis defense to a premature conclusion. They awarded me a PhD anyway.

Although I had lived and studied in the East most of my life, my parents were native westerners (indeed, I was born in Palo Alto), and I had grown to love Tucson since my arrival on that hot day in 1962. Unpleasant things were happening in Cambridge, like a convicted rapist climbing in through the window of the adjoining apartment, and then a bomb threat at the next-door building at MIT on the day of my thesis talk. So I had been looking for a postdoctoral job in the West, preferably in Tucson. Harvard's Leo Goldberg was now running Kitt Peak National Observatory at its Tucson headquarters; unlike Whipple, he looked up my record as a Harvard undergraduate and nixed me for a job there, for which two senior Kitt Peak scientists had encouraged me to apply. But my long associations with Hartmann, Alan Binder, and others helped me secure a job in autumn 1971 at what later became the Planetary Science Institute (PSI) in Tucson.

In Tucson I continued studying asteroids, with frequent trips back to MIT or to mountaintops with McCord's instrument. I published my PhD thesis research in the *Astronomical Journal* and the *Astrophysical Journal*. In a joint paper in *Icarus* (1975) with David Morrison and Ben Zellner (figure 5.5), we identified several different kinds of asteroids: what we called S-types, with absorption bands due to several silicate minerals; C-types, very black bodies with neutral, dark (carbon-black) colors; and M-types, with featureless but slightly reddish colors.[3] This began a classification "alphabet soup" that has

---

3. Clark R. Chapman, David Morrison, and Ben Zellner, "Surface Properties of Asteroids: A Synthesis of Polarimetry, Radiometry, and Spectrophotometry," *Icarus* 25 (1975): 104.

FIGURE 5.5 *From left to right*, David Morrison, Ben Zellner, and Clark Chapman, the three co-authors of "Surface Properties of Asteroids," which was published in *Icarus* in 1975 and was heavily cited in years thereafter because it introduced the taxonomy of asteroid spectral reflectance properties, beginning with the C-, S-, and M-types, related to their mineralogical composition. Courtesy of Clark R. Chapman.

expanded over the decades to use almost all letters. Most asteroids either are or resemble S- or C-types. Several colleagues were measuring laboratory reflectance spectra of meteorites; I collaborated to try to identify from which asteroid types the different meteorite types were derived. C-types were probably parent bodies for the black, carbonaceous chondritic meteorites. But S-types were puzzling. Their spectra resembled the rather rare stony-iron meteorites, but then what asteroids were parents of the most common meteorite type in museums, the ordinary chondrites? This "S-type conundrum" was widely debated in the eighties and nineties, but I became convinced that S-types were really ordinary chondrites, with colors modified by *space weathering* (changes to optical surfaces of asteroids due to bombardment by solar wind particles, micrometeorites, and so on, as had been discovered in lunar materials returned by the Apollo astronauts). Indeed I had an actual formal debate at the 1995 DPS meeting in Kona with the late Jeff Bell, who disputed asteroidal space weathering. Many witnesses to our spirited debate assumed that we were scientific enemies, so they were surprised that Jeff and I—who had long been friends—went out to dinner together afterward. My views were eventually (2011) shown to be correct by analysis of samples returned to Earth by the Japanese Hayabusa mission to the S-type near-Earth asteroid Itokawa.

My other early research at PSI included devising (in 1977, with Don Davis) the concept that many asteroids might be "rubble piles." Over the eons, asteroids collide. They fragment, but often the fragments don't reach escape velocity and instead reaccumulate into such rubble piles. The concept is now widely accepted, and in the last dozen years U.S. and Japanese spacecraft missions to several small asteroids indicate that they are, indeed, rubble piles. Brown University PhD candidate Ken Jones and I identified from his crater counts an episode of early landform degradation on Mars. I pursued studies of Mercury, and I collaborated with PSI colleagues concerning the physical nature of Saturn ring particles. As spacecraft explored the asteroids and outer planet satellites, I continued to count craters from spacecraft images to learn whether the bodies were struck by asteroids, comets, or circumplanetary projectiles, or whether many smaller craters were "secondaries" formed by fragments ejected when larger primary craters were formed. Beginning with the 1993 Galileo spacecraft images of the satellite Dactyl orbiting the main-belt asteroid Ida (figure 5.6), I began working with my colleague Bill Merline to use advanced ground-based telescopic capabilities to discover satellites of other asteroids, which had been rumored to exist since the late seventies, but it took new capabilities like adaptive optics to find them. Hundreds are now known.

FIGURE 5.6 Enhanced false-color image of the asteroid Ida and its moonlet Dactyl, to the right, taken by the Galileo imaging system in 1993. The reddish regions have been affected by space weathering, due to micrometeorite and solar wind bombardment; the bluish regions have been refreshed by recent impacts. Courtesy of NASA, JPL photo P-44131.

I won't go into such detail about my later professional career, but it clearly has been shaped, in part, by its beginnings in the fifties and sixties, at the start of the space age. I have been on the science teams of three NASA planetary missions: the Galileo mission to Jupiter, the Near Earth Asteroid Rendezvous (NEAR)–Shoemaker mission to the asteroid Eros (figure 5.7), and the MESSENGER mission, which orbited a spacecraft around Mercury from 2011 to 2015. I spent a quarter century on the Galileo mission science team; its launch was long delayed as NASA recovered from the Challenger disaster. When it finally reached Jupiter in 1996 and got good closeup images of Europa, I participated in a widely reported press conference at JPL about Europa. So I soon found myself in email

FIGURE 5.7 *Left*, Joseph Veverka (leader of the imaging and spectroscopy team) and *right*, Clark Chapman (member of Veverka's team), crouching in front of the instruments of the Near Earth Asteroid Rendezvous–Shoemaker spacecraft prior to its launch toward the asteroid Eros in 1996. Courtesy of Clark R. Chapman.

exchanges with famed sci-fi author Arthur C. Clarke. Clarke was unhappy that my first name lacked an *e*, but he was glad to see in our images the double ridges that looked to him like possible freeways on Europa.

In 1980 I had the good fortune to attend the first-ever scientific conference (organized by Gene Shoemaker in Snowmass, Colorado) on the threat to Earth from near-Earth asteroids and comets. That became a focus of my science, marrying my early interests in cratering and in asteroids. In 1989 I coauthored, with David Morrison, a trade book, *Cosmic Catastrophes*, which introduced many people, including some U.S. Congress members, to the issue of planetary defense. My earlier (1977) trade book, originally entitled *The Inner Planets* (and appearing in a later edition as *Planets of Rock and Ice*), had been favorably reviewed in the *New York Times Book Review*. My early interest in journalism and my efforts on the *ALPO Observing Manual* had trained me in how to write.

Carl Sagan, who in 1980 was organizing the Planetary Society with Bruce Murray and Lou Friedman, had invited me to be a columnist for the society's *Planetary Report*, beginning with volume 1, issue 1, and I continued to write pieces for the thin magazine for the next two decades. Under the aegis of the Planetary Society, I taught planetary science to Latin American high-school teachers in Costa Rica and in Mexico City. In March 1986 I was one of several astronomers who showed Halley's comet to passengers on a fourteen-day Cunard cruise between Rio and Miami; a quarter of the elderly passengers claimed to have first seen Halley in 1910, but many were probably recalling the Great Daylight Comet of 1910 instead.

In the mid-nineties, PSI was being reorganized in an initially unsatisfactory way, and my daughter had gone off to college, so after nearly a quarter century, I left Tucson and joined a fledgling group of planetary and solar scientists—now a large group—at Southwest Research Institute in Boulder, Colorado. I'm still there, though mostly retired and living with my wife, Y, at our ten-acre foothills residence, which Y named Rancho Europa, after Jupiter's moon. I occasionally still observe the heavens as I did as a kid. Recently I got up very early to watch comet NEOWISE and, in December 2020, the extremely close great conjunction between Jupiter and Saturn.

Back in late 2001 I was invited to a meeting at the Johnson Space Center about the asteroid threat. There I met Apollo 9 astronaut Rusty Schweickart, astronaut Ed Lu, and comet specialist Piet Hut. Together, the next year, the four of us founded the B612 Foundation, which has played a major role in informing the public, governments, and the United Nations about the asteroid threat (tiny probability, potentially enormous consequences). One day I was sitting in my living room when the telephone rang. It was Ed Lu calling from the International Space Station, telling me that he would be passing over Colorado in about twenty minutes! B612 went on to design an asteroid-survey spacecraft mission and tried to raise donations of tens of millions of dollars while NASA was largely ignoring the threat; but once NASA began promoting JPL's Near-Earth Object Surveyor mission, B612's Asteroid Institute began doing other kinds of planetary defense research.

Until the COVID-19 pandemic, my planetary career enabled me to have exceptional experiences in many exotic places in the world. My observing runs took me into domes on chilly mountaintops, though that evolved into "warm rooms" in the domes, and finally into remotely operating a telescope on Mauna Kea, Hawai'i, from my office in Boulder, Colorado. I traveled to planetary meetings around the world, in places as distant as Japan, Singapore, Australia, Uruguay, South Africa, Brazil, and Russia. In Moscow, just prior to the collapse of the Soviet Union, I was with a group of colleagues who met with the director of the USSR Space Research Institute in his fancy office. In a contrasting visit to the restroom adjacent to his office, I discovered that the toilet paper consisted of strips of the notorious *Pravda* newspaper. I did *not* take some pieces to add to my newspaper collection.

In 1990 I participated in a remarkable multiweek tour, led by Gene and Carolyn Shoemaker, of impact craters in the Australian Outback. I've been to other impact craters in Canada, Germany, South Africa, and Italy (though the Italian crater's impact origin remains speculative). And Gene Shoemaker gave me and a few others a geological field trip down into (and back up from) Arizona's Meteor Crater. I flew with Dale Cruikshank in a small plane with an open door as he photographed Mauna Kea Observatory. Since my 1954 adventure as a child, I've watched solar eclipses from the air high above Nantucket; from the tip of Baja California (Mexico); and in Narita (Japan), Jackson Hole (Wyoming), and other places. I've climbed volcanoes, one of which had erupted only days earlier, and hiked over 'a'ā and pahoehoe lavas.

Beyond my early use of what was then the world's largest telescope, the 200-inch Palomar reflector, I've used or visited many other famous telescopes, including the old Lowell Observatory telescope (another Clark refractor) and the 72-inch Rosse telescope in Ireland, which was the world's largest for more than seventy years and had just been refurbished, in part through the efforts of Patrick Moore. And I've climbed up onto the instrument platform of the Arecibo radar in Puerto Rico, which recently collapsed and destroyed the famed, enormous telescope.

I've witnessed the Saturn V launch of Apollo 14, as well as later launches from the cape of the three spacecraft missions on whose science teams I've served. I've also seen several headline-producing "UFOs," which, of course, turned out to be rocket launches viewed from a great distance. The first was a strange comet-like thing low in the southeast while I was observing from my backyard in Buffalo; there were UFO headlines in the Buffalo newspapers, but I explained that it was the first Tiros weather satellite launched northeast from Florida. Another "UFO" was viewed straight down the tail from northwest Tucson of a Vandenberg launch six hundred miles away. It created banner headlines in the Tucson newspapers about strange flying saucers and little green men in my neighborhood.

Despite deep public fascination with space exploration, astronomy and especially planetary science are actually very small fields of researchers, unlike chemistry, physics, or biology. So I've gotten to meet and know many in my field over the years. As a freshman, I was in a small group that spent an evening with Harlow Shapley. In the sixties and early seventies, I probably knew almost every American planetary scientist, and I got to know and work with many from other countries in later decades. The field all started with the "founding fathers," like Kuiper and Dollfus, and now has many bright young graduate students just getting started.

One event in 1997 brought things full circle for me. I had become friends with David Levy, the Canadian amateur astronomer who, after moving to Tucson, became a leading discoverer of comets using several telescopes in his "telescope farm." He famously discovered, along with Gene and Carolyn Shoemaker, comet Shoemaker-Levy 9, shortly after it had been tidally disrupted by Jupiter. Its fragments then crashed spectacularly into the giant planet during a week in July 1994. I had led the Galileo spacecraft's observations of the comet crash from its oblique orientation, with a direct view of ground zero, which was hidden from Earth on Jupiter's backside. In July 1996 David proposed to Wendee on the Eiffel Tower, and then three days later they joined me, my wife, and the Shoemakers on a Bateaux Mouches dinner cruise on the Seine. In March 1997 David and Wendee were married, and we joined a crowd of their family and friends at their home in the Tucson exurb of Vail for a wonderful reception. It was a clear evening with a near-total eclipse of the Moon under way and with comet Hale-Bopp gleaming in the northwest. My wife, Y, wanted to see the comet through one of David's telescopes, and we waited for the man at the eyepiece to climb down. When he did, I introduced Y to comet-discoverer Tom Bopp. Later she would say that she immediately thought, "Yes, and the next person will be Santa Claus himself!" The reception at the Levys was a lively event. Even Walter Haas danced in the hora line, though a bit awkwardly. It was the last time I saw Walter, who was so instrumental in his encouragement of my interest in planets, beginning only a couple years after Sputnik was launched.

What is the future for exploration of space and the solar system? Just a few months before Sputnik, the Kiplinger magazine *Changing Times*, which my parents subscribed to, had an issue devoted to predictions for life twenty-five years hence, by experts from all segments of society. Reflecting the optimism of the times about going into space, I recollect that predictions included colonization of Mars not as soon as twenty-five more years but definitely before the end of the twentieth century. Nonetheless, predictions of the future are notoriously difficult. The experts in 1957 hardly foresaw the electronics revolution, though computers, smartphones, the internet, and so on were already under way by the early eighties and now dominate the lives of many people. Nowadays, rich entrepreneurs like Elon Musk, Richard Branson, and Jeff Bezos, and countries including China, are forging the way into space. Yet on the other hand, despite hopes that we may have left the disastrous political and social forces of the last century behind, it isn't looking good as I write this, when the future of democracy nationally and globally is under threat. There is an extremely tiny chance that an asteroid or comet impact will doom civilization, but the hopes and fears of humankind almost certainly lie with the actions of those of us here on planet Earth.

6

A Golden Age Revisited

WILLIAM SHEEHAN

We lived through a golden age of Solar System astronomy and saw the birth of planetary science as a discipline.

—William K. Hartmann to William Sheehan, Dec. 22, 2020

I was born on what would be the center line of the total solar eclipse of June 30, 1954, though since the eclipse occurred only twelve days later, it made no impression on me whatever. My earliest possible astronomical memory is of being taken outside to see Sputnik in October 1957. In all likelihood it is a false memory, one implanted in me by what the adults around me said later, though it is not impossible that it records an actual experience. Be that as it may, it signifies that by the age of three or so, I was at least vaguely aware that the space age had begun.

Growing up in the United States during the time (after World War II) when it had over half of the world's GDP (it has less than a fifth now), I took for granted that ours was the exceptional, the indispensable country. We were America the beautiful, where blessings flowed from sea to sea. We welcomed immigrants and new ideas. We were scientifically and technologically advanced. We were committed by our Constitution to the values of the Enlightenment, which included reason, liberty, tolerance, and a commitment to progress. It was easy to take for granted that progress was inevitable and that the Enlightenment, once achieved, was irrevocable. Within a few years, the civil rights and women's rights movements began to highlight our flaws, and the Vietnam War and its aftermath would break the country into a million pieces. Even now, fifty years later, it hasn't been put back together again.

My father, a factory worker, had only an eighth-grade education, and my mother was a homemaker, a high-school dropout who had found little nurture in the public schools of the forties (which assumed that men were the ones deserving of a good education and that women would just be charged with child rearing anyway, so what was the use?). We had very few interesting books around the house, though since I grew up a stone's throw from a public library branch, this didn't much matter. I attended a school where the teachers were Benedictine nuns, and one or two of them were very interested in the natural world (birdwatching, horticulture) and science, and they encouraged—or at least did not discourage—my developing interest. Though times have changed, and there are no longer enough nuns to support such schools on the scale on which they existed then, I have nothing but admiration for these dedicated individuals, who were swamped by the baby boom (more than forty students in a class) and probably undereducated for the tasks they were expected to carry out. I must mention one of the nuns in particular, Sister Esther, who started the school's first science fair and shared my dreams (though she never saw them fulfilled) of building and using telescopes to observe the Moon and planets. I will always have a special place in my heart for her.

Though I was born and raised in Minneapolis, my grandmother had grown up on the Iron Range, where her father had been a miner, and every summer I was taken there to visit the relatives on that side of the family. I was fascinated for the first time by something geological—the rust-iron hills, which during World War II had provided the raw material for the steel that industrialized America and helped it win the war against Germany and Japan. While pondering the enormous piles of tailings from the open-pit mines, I was also alerted to geological signposts, which told of events going back not thousands of years, like those that encompassed all the modern, medieval, and ancient history I had known about (the centennial of the Civil War was celebrated when I was seven, and that seemed like a long time ago), but billions of years. Thus began a slowly dawning realization that perhaps what I had been taught wasn't the whole story, and that a vast stretch of time, shadowy and unknown, lay beyond the brief flicker of our (*Homo sapiens'*) time on Earth. In due course, this led to an even greater and more wonderful discovery—that of the vastness of space, and of other worlds beyond the one we knew.

During this early period, the space program was getting under way, and I took at least a broad interest in it. For instance, though I mostly followed the American developments, I was aware that a Russian (I did not know his name) had been the first man in space. I still remember vividly getting up early on the several occasions when John Glenn's launch was scrubbed before finally seeing him successfully off on his historic three-orbit flight on February 20, 1962, and the pride I felt. I regarded him as a real-life hero, in contrast to those I saw on TV, such as Superman and the Lone Ranger. John F. Kennedy was another hero; I remember hiding under desks at school during the Cuban missile crisis, and being terrified by the prospect of nuclear war with the Soviets, but confident in his leadership. Though the Cuban missile crisis overshadowed Kennedy's address at Rice University (September 12, 1962), where he said, "We choose to go to the Moon," I became gradually more and more aware of our aspirations in space—and had more time and energy to devote to it after I failed disastrously at baseball, which, with Civil War history, had been my main interest in first, second, and third grades.

But what I have said so far is prologue. My real interest in astronomy began—and I know not only the day, but the hour—on February 17, 1964, when I was nine years old. It was a Monday, and I was on my way home from school. Though it was still winter, the temperature in Minneapolis reached an unusual forty-five degrees Fahrenheit. The snow melted and ponded into puddles, and I couldn't resist splashing through the stands of water in my overshoes. Perhaps the warmth and brightness of that day made me more distinctly conscious of the Sun than I would have otherwise been: of its round perfect disk reflected in these pools of water, the dramatic circles of light breaking, as if by magic, into electric dots and dashes—codes of wonder—as I splashed through them. For whatever reason, something occurred to me in a flash of awareness. I realized, as Giordano Bruno and Nicholas of Cusa (and perhaps Democritus) had done, that the Sun is a star, the stars, distant suns. I was awestruck and amazed. I had blundered into something that was hardly due to mere chance. It would amount to the opening of a destiny.

I ran all the way home and immediately looked up every word I could think of related to astronomy in the dictionary: Sun, Moon, planet, star. I observed the Sun through my bedroom window, staring at it so long and hard that I experienced a case of solar retinitis. (My enthusiasm obviously far outpaced my knowledge or discretion, and I had not yet read the oft-repeated warning in astronomy books: NEVER LOOK AT THE SUN WITHOUT PROTECTION FOR THE EYES.) Next I looked long and hard at the Moon, a three-day-old crescent hanging among the branches of an elm tree in the backyard.

At this time, Venus was appearing as the evening star, slowly working its way toward its greatest elongation east of the Sun, which it passed in mid-April. It became a compelling pastime for me to find out how early I could first glimpse its sharp pinprick against the blue-gray sky, and before long, my young eyes (recovered from retinitis) were able to make it out in broad daylight. I wondered at this lovely planet, and its beautiful lucid clear light, with an intensity of interest that no Sumer-Akkadian or Mesoamerican Venus worshipper could have surpassed. I cannot count the number of times I wished myself far from my troubles in this world into that seemingly perfect, eternal, and untroubled orb of light.

Anthropologists tell us that nearly all societies (including ours) engage in practices that lead to altered states of consciousness. Some of them have a strong belief in possession of the body by spirits, or in the soul fleeing the body to undertake an ecstatic journey. For example, Maria Nikolaevna, the mother of Sergei Korolev, the chief designer of the Russian space program, recalled that as a child, he had liked stories inspired by the Tunguso-Manchurian shamans, in which "we 'flew' together on a fairy-tale magic carpet and saw Korek-Gorbunek (the humpback pony), the grey wolf, and many other miracles beneath us. This was enthralling and wonderful. My son would press close to me, would look wide-eyed into the sky, while the silvery moon peeked out from amidst the small clouds."[1]

Through astronomy, I too achieved an altered state of consciousness. I found an escape from the rote drudgery of schoolwork, the threat of bullies lurking in narrow alleyways, and the unpredictable moods or capricious rules of adults. Standing in my small backyard, looking up at what Homer called "the broad heaven," I was a being set apart; all my troubles, all the rest of the world, fell away; I was oblivious to them and found myself transported above the clouds and into the wondrous realm of the Sun, Moon, planets, and stars.

Though I couldn't afford to buy many books then, I discovered at the library a small shelf entirely devoted to astronomy, which I immediately claimed as my own. No one ever seemed to challenge me for it.

Many of the titles were by Patrick Moore—*Story of Astronomy*, *Guide to the Moon*, *Guide to Mars*, *The Planet Venus*, and (with H. P. Wilkins) *How to Make and Use a Telescope*. He was a one-off, the man for the moment. Com-

---

1. Quoted in James Harford, *Korolev: How One Man Masterminded the Soviet Drive to Beat America to the Moon* (New York: Wiley, 1997), 17.

ing on the scene just as the space age was beginning, and when relatively little was definitely known about the Moon and planets, he brought an incredible enthusiasm and the literary equivalent of perfect musical pitch—the ability to aim his writing precisely at the point where an intelligent twelve-year-old could understand it perfectly. As an amateur astronomer himself, he promoted the idea, which had long been advanced by the British Astronomical Association, that there was still a great deal of useful work to be done even by someone with only a small telescope and limited artistic skill. It helped, of course, that he also began to broadcast a monthly TV program, *The Sky at Night*, just as a bright comet (Arend-Roland) was appearing in the sky and as Sputnik went into orbit.

In Patrick's books (and others), I read with keen interest stories now so worn into familiarity that the discovery of any new detail is an event. But then, I encountered them for the first time. Ptolemy, Copernicus, and Tycho Brahe became heroes, and all did as I did—observe the planets with the naked eye. Before long, however, I longed to have my own telescope and somehow managed to scrape together thirty dollars—it was a large sum for me at a time when my weekly allowance was a quarter—for a 60 mm refractor, one of the much-despised department store telescopes, but for me, a source of countless hours of pleasure and instruction.

I will never forget the first night I used it—Saturday, September 12, 1964. Naturally, I aimed for the Moon. I set up the telescope in my grandmother's front yard, beneath oak trees, and pointed it at the five-day-old Moon then standing among the stars of Scorpio. Putting my eye to the eyepiece, I saw an image that was sharp, chiseled, strange: a battered face of craters, plains, mountain ridges, and other things I did not recognize. Gone was the Moon as it had hitherto been, round and small enough to catch within the palm of my hand; gone, too, the homely face of the man in the Moon. All of this fell away, and something wonderful appeared. Instead of seeing the lunar globe from a quarter of a million miles away, the image, magnified thirty-five times, showed it as it would look from the porthole of a spacecraft from only a few thousand miles away. Next I "discovered" the satellites of Jupiter, the phases of Venus, the ring of Saturn. Though Hubble Space Telescope observations indicate that more than one hundred billion planets are likely in the galaxy, the sight of none of them could ever give me as much satisfaction as those nearby planets seen through my small telescope in the winter of 1964–65. One has that experience but once in a lifetime.

FIGURE 6.1 Apologies to Ray Walston, "My Favorite Martian." The author standing in front of the garage door, a year before Mariner 4's flyby. Courtesy of William Sheehan.

Mars, however, was then and long afterward my main interest, because of the possibility of life of some kind, and even intelligent life, there. I had been captivated by Percival Lowell's ideas of a dying planet, whose inhabitants had built a network of canals to stave off encroaching—and ultimately inevitable—doom. For a year or so after I became so passionately interested in astronomy, it was possible—just—to believe that Lowell's vision might be reality (figure 6.1).

I first looked at Mars with my telescope in March 1965, when it came to an unfavorable (aphelic) opposition. The opposition date was March 9. (To put this in context, it was the day after the first of two U.S. Marine battalions started coming ashore at Da Nang in South Vietnam. By the following summer, it would be impossible even for someone of my tender age not to appreciate that the country was heading, step by step, into a great catastrophe in a part of the world we understood perhaps even less than we did Mars.) The disk was too small to show much detail, but even so, I found that warm, ruddy little pin's head of light ever so beautiful and studied it intently for hours. The artist John Ruskin once said, "The greatest thing a human soul ever does in this world is to *see* something, and tell what it *saw* in a plain way. Hundreds of people can talk for one who can think, but thousands think for one who can see. To see clearly is poetry, prophecy and religion, all in one."[2] Little as it was, Mars I had *seen*, and in a plain way. There was poetry, prophecy, and religion in that.

Even then, NASA's Mariner 4 probe was three months out from Earth on its way to Mars, due to arrive at the red planet on July 14, 1965. I could hardly wait. Eventually the day came. The camera system sent back bits of each picture in packets that corresponded to the brightness of each pixel of an image. (The word *pixel*, by the way, was first published that same year by a

---

2. John Ruskin, *Modern Painters*, vol. 3, part 4, chapter 16, section 28 (London, 1856).

Jet Propulsion Laboratory engineer to describe the picture elements of such scanned images.) When they were finally released, the Mariner 4 images were not very clear, but they revealed a stark reality: no canals, only craters, impact craters, something like three hundred in all, in all sizes, including one in Mare Sirenum measuring 120 kilometers across. It was as if Mariner had simply photographed a strip of the Moon's surface, with a thin hazy atmosphere superimposed. The photos were released for some time. Those days of Mariner 4's sobering revelations were among the saddest of my life. Mariner 4 had killed my fond Lowellian dream and taught me a hard lesson: reality is no respecter of wishful thinking.

Well, I eventually got over my disappointment. There were, in any case, other worlds than Mars (and Mars had much more to offer than seemed at the time). Perhaps there were new worlds not even known. I read a book called *The Search for Planet X*, by Tony Simon, that described the Lowell Observatory's long quest for a planet beyond Neptune and Clyde Tombaugh's discovery of Pluto. This fired my imagination, and soon afterward, I found a tiny speck moving back and forth near Delta Geminorum on two plates reproduced in James Jeans's *The Universe Around Us*. Could it be a planet beyond Pluto? I wrote to Lowell Observatory and the secretary there, Helen Horstmann, through whose hands every piece of correspondence going into and out of the observatory passed and who made it a point to answer queries from children. She no doubt gave it to Robert Burnham Jr. to answer. Burnham was comparing Clyde's plates with more modern ones to detect stars of large proper motion, so he was in a good position to respond to me. He pointed out that whatever I had found, it was not a planet—probably it was a speck of dust or a random aggregation of silver halide grains in the plate. But rather than end on a discouraging note, he provided me with information about a comet discovered by two Japanese observers, Ikeya and Seki, that belonged to the family of sun grazers and might become brilliant during its nearest approach to the Sun at the end of October 1965. Though there were no guarantees, since comets were notoriously unpredictable, he provided a few suggestions on how to look for what would become one of the great comets of the twentieth century, and which, all thanks to the kindness of a stranger, I saw on several mornings in early November. The head, which had been spectacularly brilliant when the comet approached perihelion, had been then faded rapidly and was no longer visible with the naked eye, but the enormous ghostly tail was still easily visible, streaming across the sky like a searchlight. I was filled with awe, which at that time was still a largely religious response for me.

I continued to pursue my astronomy, with a larger telescope (a 4.25-in. Edmund Scientific Palomar Junior), which showed me enough detail to make sketching the changing clouds of Jupiter worthwhile but wasn't quite large enough to satisfy. Gearing up for the junior high science fair, I began grinding a 10-inch mirror in the basement. (As the saying goes, my eyes were bigger than my stomach, and it was never finished.)

One of the themes that emerges from the other chapters in this book is how many of those who went on to professional careers in planetary science not only demonstrated precocious aptitudes for science (and often representational art), but also came from families steeped in physics, engineering, chemistry, pharmacology, or other technical subjects. Many of them also found adult mentors in astronomy, who not only stimulated their interest but, at crucial times, provided them with letters of recommendation or other encouragement that would swing doors open to advancement in their fields. I had a precocious but almost entirely self-directed interest in astronomy and very little in the way of adult mentoring (if I had had that, I would doubtless have finished the 10-inch telescope and become an avid amateur observer of the planets, especially Mars and Jupiter, much earlier than was actually the case; the 4.25-inch telescope was too small to do more than whet my appetite for more). Consequently, I was much slower in developing than many of the others whose autobiographies appear here.

The high school I attended (Patrick Henry in North Minneapolis) reflected the lower middle-class neighborhood in which I grew up, and in which I and others experienced a good deal of deprivation, mostly having to do with a lack of resources, encouragement, or expectations—though, I hasten to add, nowhere near the sort of disadvantages associated with inner city youth. It was not very strong in math or science (or anything else). Unchallenged in academics, I proceeded to spend much of my time and effort in athletics, and, surprising myself, became quite an accomplished miler, at least by the standards of the school and time. Those were the years of the Apollo Moon landings—all except Apollo 17 took place while I was in high school. I was also more steeped in running than anyone else when the long-awaited perihelic Mars opposition of August 1971 took place, which I observed with the 4.25-inch reflector. Modest though the means were, they were sufficient for me to record the progress of the global dust storm that began in late September and was still very much in evidence when Mariner 9 went into orbit around Mars that November. As the dust gradually began to clear, and the Tharsis shield volcanoes, canyons, and dry riverbeds came into view, I followed developments as best I could from *Sky & Telescope*, to which I had begun to subscribe, though my academic progress was more along literary than scientific lines, as I encountered, during my last semester of high school, the first really exceptional teacher I had ever had: Doreen Savage, who taught a course in the novel and sparked a love of literature, which soon became almost all-consuming. Thanks to her, I was soon writing scads of poetry, including epistles in heroic couplets and a blank verse tragedy, and she encouraged my "way with words" and a belief that I might actually become a poet, my path similar to that of W. H. Auden, who started out interested in science and engineering, which led to a scholarship at Oxford, but switched his field to English after becoming fascinated with poetry. Doreen was the first adult mentor I found, though for literature rather than science, and I owe to her a great deal of the success I had subsequently.

Despite my family's modest means , there was never any real doubt that I would attend college, and I seriously considered only the University of Minnesota, a largely commuter land-grant school (it was the only school I considered that didn't expect me to continue my running career). Despite my sudden enthusiasm for poetry, I decided to major in physics. My adviser was Alfred O. C. Nier, who was famous for his work on the mass spectrograph and was helping design the ones used by the Viking landers on Mars. The Nier Prize, awarded annually by the Meteoritical Society for outstanding research in meteoritics, is named after him. But he wasn't very interested in undergraduates; he met with me for all of five minutes, signed off on the courses I wanted to take, told me there's always room at the top, so that if I was good enough, I would make my way forward as he had done, and there was no need for me to meet with him again. Of course, he had absolutely no reason to take an interest in me at that stage (figure 6.2).

I took all the math and physics courses I could, and though I was still hugely interested in astronomy, the offerings for undergraduates were scant. I had an excellent general course on astrophysics taught by Jay Gallagher, then a recently minted PhD with long hair, who wore flower-print shirts and bell-bottoms; he subsequently went to Wisconsin and had a brilliant career researching star formation in galaxies, dwarf galaxies, and dark matter. The only upper-division courses on offer, however, were Astrophysics of Diffuse

FIGURE 6.2 Harrison (Jack) Schmitt on the Moon, December 1972. Painting by William Sheehan based on an Apollo 17 photograph by Eugene Cernan. The author had just finished his first quarter as a physics major at the University of Minnesota at the time and never dreamed that this would be the last manned lunar landing for more than fifty years. Courtesy of William Sheehan.

Matter and Astrophysics of Condensed Matter, titles that didn't exactly fire my imagination and were usually reserved for graduate students. I never took them. Indeed, astrophysics was rather closed off for all but the most advanced students, and though Ed Ney and his group were doing some excellent research in infrared astronomy, his attitude toward undergraduates was the same as Nier's. Any thought I might have had for a career in astronomy and astrophysics was forestalled. I never even considered pursuing coursework in subjects like geology, meteorology, and the like, or a career in planetary science; it was simply not done at that time or place. What I did, however, was continue to observe the Moon and planets, by now—since my 60 mm refractor and 4.25-inch reflector had been stolen from my garage in my senior year of high school—using a 6-inch Dynascope reflector belonging to Mike Conley, a friend to whom I had introduced astronomy in high school and who has continued to be a good friend, sharing that interest to this day, fifty years later. During the fall of 1975, Mike and I spent entire nights observing and sketching Jupiter during the period when the planet was experiencing a rare triple-sourced south equatorial belt disturbance, and its clouds were roiled into a bewildering array of intricate and ever-changing forms. The understanding of these cyclical Jovian phenomena was, by the way, almost entirely owing to amateurs, including Association of Lunar and Planetary Observers Elmer J. Reese and Clark Chapman.

Some good people in physics were willing to mentor me, most notably Alan Goldman, a brilliant low-temperature physicist who had been narrowly beaten into publication (and hence to a Nobel prize) for the discovery of what is now known as the Josephson effect, who invited me to hang out in his lab in the basement of the physics building. But I was still trying to master classical physics, so low-temperature physics, which was all about quantum effects, left me rather cold. Meanwhile, the prospects of a research career in the physical sciences were being overtaken by events: runaway inflation associated with the Vietnam War led to the Nixon cutbacks in the federal budget, which the OPEC oil embargo made worse, and there were no longer so many opportunities for careers in space as had existed during the era of "budgetless financing." Because of tenure, careers in teaching were even harder to come by.

Thus, despite completing the coursework for my physics degree in three years, I decided there was no rush to get on with things and "hung fire" for a while, mostly taking courses in the English Department. I eventually finished the coursework for an English degree, and so I became the rather anomalous-seeming physics and English lit. double major and was encouraged to "develop broadly." I combined elements of John Keats's poetry with Niels Bohr's physics in an honors thesis that delighted my professors, and my success led me to apply to a good interdisciplinary program, where I could continue to straddle C. P. Snow's famous "two cultures' gap." One of the few such programs in existence was at the University of Chicago. I applied, and the chair of the department wrote back: "I have never yet seen an application which had the special range of endowment and humanistic concern that yours evinces." So in the fall of 1977, as recipient of a Special Humanities Fellowship to pursue a year of general studies in the humanities, I got on the Greyhound bus in Minneapolis—it was all I could afford—and went to Chicago.

Nowadays, *interdisciplinary* is the latest buzzword, but in 1978, such programs were very rare, at least in the humanities (they were beginning to become common in science). Though the idea behind them may have been worthwhile, in practice, such programs were rather difficult to navigate. The academic system itself at the time—with its departmental silos, or as they are now being called, stovepipes—had tremendous inertia to continue doing things the way it already knew. I was clearly a big picture kind of person, a generalist, but what I had begun to realize was that a generalist could not really compete with the specialists, at least not in the short term. Some of the courses I took—especially a seminar, Imagination/Representation, taught by world-renowned philosophers Paul Ricoeur and Stephen Toulmin, which started by assigning Kant to be read in the original German—were nearly incomprehensible, or at least laughably far over my head, and at the end of a year, I still had no idea in what I might specialize in for a PhD. My mentors suggested I might opt for English, history of culture, history of science (a subprogram in the History Department), conceptual foundations of science, or the Committee on Social Thought. None of these options seemed appealing. Painful as it was to forfeit the fellowship, I felt I had reached a dead end and decided to call it quits with pursuing a master's degree in general studies in the humanities.

I was pretty depressed and trying to sort all this out when, in May 1978, University of Chicago alumnus Carl Sagan came to the campus. He was already well known and represented one of the truly exciting fields of interdisciplinary studies, which had been developing right under my nose: planetary science, which combined astronomy, geodesy, cartography, geology, geophysics, meteorology, cosmochemistry, even astrobiology. Sagan was returning to Chicago for the first time since he had been a student there to give a talk in Ryerson Hall on the latest Mars results from the Mariner 9 and Viking missions. I arrived early and sat on the floor ahead of the first row of chairs; I was literally sitting at the master's feet. Despite my early passion for Mars, I had struggled to stay abreast of all the exciting discoveries being made, and Sagan's lecture immediately rekindled my interest in planetary science—especially Mars. The idea of actually becoming a Mars researcher, however, seemed impracticable. I didn't have the specialized knowledge. Instead, I thought of trying to teach science, at a secondary school level, and perhaps write about it for the general public. Nothing came of any of that at the time. Yet, during my year of taking courses in secondary school education, including student teaching, I met my future wife, Debb Nelson, who was a TA in a required course (aptly called Human Relations). We married in December 1980.

While still in high school, I had started writing the manuscript of a book on planets, "Worlds in the Sky," which I had set aside for several years but now took up again as part of my developing program of trying to communicate the excitement of science to the wider public. Sagan was, of course, about to do just that with his *Cosmos* book and TV series. My role model was still Patrick Moore. But when Patrick, who was always an amateur, wrote, the Moon and planets were still very much the purview of amateurs, and planetary science as a discipline had hardly developed. Now so much was known that anyone attempting to write in the field had to have a great deal of professional training. Instead of Moore (and before Sagan came to dominate the field), some of the best science writing I knew was by Clark Chapman (author of one of the chapters in this book). His *Inner Planets* (1977) made a strong impression on me and showed me that "Worlds in the Sky" was not by any measure ready for prime time. Meanwhile, in 1980–81, Voyagers 1 and 2 flew past Jupiter and Saturn—I remember following Voyager Imaging Team leader Brad Smith's press conferences and being utterly enthralled by all the discoveries about those hitherto remote and inaccessible worlds. In light of what was being found out, I completely revised "Worlds in the Sky" (and little dreamed then—and could not even have imagined—that a few years later, the book would be published by the University of Arizona Press, and that Brad himself would write a positive review of it in *Sky & Telescope*).

Meanwhile, in the late summer of 1981, I was hired to teach physics, math, and astronomy at a community college, and in that role I penned an article for the local astronomy club's newsletter, summarizing some of the factors involved in visual observations of the Moon and planets. Despite the obscurity of this publication, somehow a *Sky & Telescope* assistant editor named Stephen James O'Meara came across it. It intrigued him, and he wrote to me to describe his own philosophy of visual planetary observing. A few years earlier he had discovered, using the 9-inch refractor at Harvard College Observatory (though not a Harvard student himself), a series of faint spoke-like markings on the B ring of Saturn, a discovery that would soon be confirmed by Voyager 1.

Steve and I later became good friends and fellow observers, and his letter reawakened a dormant interest in how earlier observers had gotten things so wrong about the Moon and planets—Percival Lowell, of course, with the canals of Mars that had so fascinated me as a child, but also Giovanni Schiaparelli with his observations of Mercury (which two of the authors in this book, Dale Cruikshank and Clark Chapman, had written about) and others. I thought it might have to do with the way the eye-brain-hand system works under challenging conditions, like trying to make out details on a small disk through the medium of a continually disturbed atmosphere, though I didn't really know very much about the psychology of perception at the time. I also appreciated that it would be useful to examine not just the published results but the original observing records of people like Lowell and Schiaparelli, and so—rather boldly in retrospect—I wrote to the Lowell Observatory to find out whether this would be possible. Somewhat surprisingly, I received a very encouraging response from the Lowell director, Art Hoag, who invited me to spend some time at Lowell to do research in the archives. Debb and I spent some time there in the summer of 1982.

I must briefly digress to mention that, though most of my interactions were with Art and Lowell historian Bill Hoyt, I also met Heidi Hammel, then a summer student from MIT staying in the historic (and soon to be demolished) Vesto M. Slipher house. I remember only that she was an excellent guitar player. I could not know then that, after doing her PhD under Dale Cruikshank (one of the contributors to this book) at the Institute for Astronomy, University of Hawai'i, and becoming fascinated with the still largely unknown outer worlds Uranus and Neptune, she would go on to become one of the leading planetary scientists of her generation—notably serving on the Voyager 2 imaging team during its Neptune flyby in August 1989 and famously leading the Hubble Space Telescope team, which captured dramatic images of the likely never-to-be-repeated-in-our lifetime impacts of Shoemaker-Levy 9 comet fragments onto Jupiter in July 1994. More recently she has played a leading administrative role in astronomy, serving as executive vice president of the Association of Universities for Research in Astronomy (AURA). She is a great astronomer and leader, and I am happy to say that I knew her when.

Most of the supervision I received while at the observatory was provided not by any of the astronomers but by Bill Hoyt, a former newspaperman who had produced seminal and still valuable histories of Lowell Observatory, including *Lowell and Mars* and *Planets X and Pluto*. He was interested in some of the same questions about visual planetary observations as I but didn't have any personal experience actually observing and sketching planets, so he tended to defer to me on those questions—which was extremely gratifying to my ego at the time. He turned over the keys to the vault, which contained,

FIGURE 6.3 At Lowell, July 1982: *left to right*, Mike Crowe of the University of Notre Dame; Bill Hoyt, historian of Lowell Observatory; and Art Hoag, director of Lowell Observatory, standing in front of a portrait of the observatory's founder, Percival Lowell. Courtesy of William Sheehan.

in somewhat disorderly condition, a "treasure trove" within, of which so far the surface had hardly been scratched (figure 6.3).

Art was not only a brilliant research astronomer and effective administrator, but also one of the kindest and most encouraging people I've ever known. He generously gave me the keys to the dome of the 24-inch Clark refractor (something that would not, alas, be possible today, since the telescope is completely subscribed for the public observing programs), and I set out to make my own observations with that storied telescope, for comparison with the sketches I was examining in the observing logbooks (figure 6.4). Three planets were then visible in the evening sky—Mars, very far away, and Jupiter and Saturn. Every evening I would traipse up from the apartment in the Administration Building, where Debb and I were staying, to the dome located near Percival Lowell's mausoleum, open the shutter, and follow the planets as they sloped down the western sky into the Douglas firs and ponderosa pines. What a magical time!

Somehow the juxtaposition of planet sketches in observing logbooks and direct observations through the telescope produced a strange alchemy

FIGURE 6.4 The author observing with the Clark refractor at Lowell Observatory, July 1982. Courtesy of Deborah Sheehan.

in my mind and led to a eureka moment, comparable to the one I had experienced as a young child traipsing through puddles and suddenly being struck with the realization that the Sun was a star. It remains to this day one of the "spots of time" the poet Wordsworth once wrote about.

As enthralled as I had once been by Lowell and his theories of intelligent life on Mars, I had for years after they had been debunked by Mariner 4 paid close attention to the rather extensive literature on the subject—reading not only Hoyt's book but the manuscript of *The Extraterrestrial Life Debate*, which Professor Mike Crowe (a correspondent by now) was then writing. These works discussed in exhaustive detail the efforts of nineteenth- and early twentieth-century astronomers to explain the canals in psychological terms as illusions. Mostly they assumed that various splotches and disconnected details of the Martian surface, just at the limit of resolution, were strung together by the eye and brain into continuous lines. I had known of that theory since I had first encountered it in pre–Mariner 4 days, and it was convincing—up to a point. I sensed that something had been missed. In particular, from my firsthand experience of observing planets, I knew that planetary surface details did not hold steady like the fruit in a bowl or flowers in a vase represented in still-life paintings. The atmosphere streamed like a river in front of the planet, and the image more closely resembled the flickering series of a motion picture in which the camera is mostly out of focus but now and again stands out magically sharp, in what Lowell called "revelation peeps." In such moments, one might exclaim, recalling Hamlet, "'Tis here! 'Tis there! 'Tis gone!"

I knew there were psychological experiments having to do with temporally challenged perceptions but didn't remember what the instruments used for them were called. I was pursuing this line of inquiry with Debb in Cline Library on the campus of Northern Arizona University and asked her about it. She had studied psychology and remembered at once: tachistoscopes. That moment was another "spot of time" never to be forgotten, and the following spring of 1983—while she was working as a nursing home administrator, and I had nothing to do but write while we lived (bided our time) in a small town called Leroy, Minnesota (population 800), near the Iowa border, with no research facilities other than a Carnegie library with just one six-foot shelf of books on math and science—I wrote at white heat the manuscript of a book that, surprisingly, would become rather famous: *Planets and Perception.*[3]

We had rusticated ourselves in that rather bucolic location because just then the United States was experiencing the worst economic downturn since the Great Depression, triggered by tight monetary policy from the Fed in an effort to fight mounting inflation. Autobiographical accounts of astronomers often fail to consider the wider social and economic contexts that as much as anything determine the direction of careers—and lives. I had no idea then that my exploration of visual planetary astronomy would prove to be anything more than a lark, but I wrote the book at white heat because by then, my time to work on it was up against a self-imposed deadline—for, lacking any better idea, I had decided to apply to medical school and had been accepted to begin my medical studies at the University of Minnesota in the fall of 1983. There, I would have no time to work on the history of planetary studies. The manuscript, completed except for the last chapter, was thrown into a desk drawer and more or less forgotten while I became absorbed in learning nerves and muscle insertions of the human body in anatomy, interpreting cross-sections of tissue seen under the microscope in histology and pathology, mindlessly committing to memory the steps in glycolysis and the Krebs cycle in biochemistry, distinguishing the characteristics of the various microbes that throughout human history had ravaged humanity in epidemics and pandemics (this was interesting!), and all the rest of the enormous body of trivia that a medical student needs to master—or at least hold in memory—until after the next exam, at which point it could be promptly forgotten.

Though little of what I learned in med school was relevant to my sideline researches, I did find interesting the lectures on visual physiology by Richard Purple (a student at Rockefeller University under Keffer Hartline, who won

3. William Sheehan, *Planets and Perception* (Tucson: University of Arizona Press, 1988).

a Nobel Prize for demonstrating lateral inhibition in the retina of the horse-shoe crab *Limulus*), and I included some of what I learned from Purple in the last chapter of my manuscript, whose completion meant I could finally begin to submit it to publishers. No one expressed any interest in it except the University of Arizona Press, which had published Bill Hoyt's books. After submitting it, I heard nothing for the longest time (though at least that was better than a prompt rejection letter). At last, in the fall of 1986, the press's acquiring editor, Barb Beatty, called while I was somewhere on the East Coast visiting possible residency sites (since I had always been interested in the brain, and in acquiring knowledge about the nature of human knowing, I never considered any other specialty except psychiatry). She expressed an interest in sending it out for academic review. Since my manuscript was an interdisciplinary work (what else?) involving both history of astronomy and perceptual psychiatry, she recognized that it was unlikely that she could find reviewers with expertise in both areas, so in the end, it went out to three reviewers, two astronomers (Ewen Whitaker and Ron Miller, as I learned later) and one psychologist, whose identity I never found out. They were all effusive in their praise, with the psychologist neatly summing up my thesis about visual planetary observations: "Once a definite expectation is estab-lished, it is inevitable that one will see something of what one expects; this reinforces and refines one's expectations in a continuing process until finally one is seeing an exact and detailed—but ultimately fictitious—picture."

So the book came to be published in the fall of 1988. By then I had a son and was well along in the internship year of my psychiatric residency (served at the Veterans Administration Hospital in Minnesota). A first review ap-peared in November, penned by the well-known British amateur astronomer and historian Richard Baum, in the *Journal of the British Astronomical As-sociation.* I was on the neurology service and read it one night in the on-call room at the VA; it was very favorable, though knowing Richard as I came to know him, it could hardly have been otherwise. I expected at the time that it would likely be the only review I ever got. In May 1989, however, when I was back on the psychiatry service, the attending physician, Charlie Dean, who subscribed to *Nature,* casually mentioned at the end of morning rounds, "Congratulations, by the way, for the nice review." The review was by Albert Van Helden of Rice University, renowned as the world's leading authority on the history of the telescope, and it came out in one of the leading journals of science in the world, *Nature.* Dean must have told the chief of psychiatry,

Richard Magraw (something of a renaissance man, whose book *Ferment in Medicine*, which anticipated some changes in medicine that were about to occur, such as the introduction of managed care, had been a bestseller in the sixties); Magraw greeted me in the hallway as I was headed to my outpatient medicine clinic with the words from Coleridge's "Kublai Khan":

> Weave a circle round him thrice,
> And close your eyes with holy dread
> For he on honey-dew hath fed,
> And drunk the milk of Paradise.

Old Sister Esther would have been proud of me at that moment.

To make a long story short, Van Helden's review was followed by many others. In another chapter in this book, Chapman notes just how small the circle of planetary scientists is, and that is what I found out as well. Appreciative reviewers included such eminent planetary scientists as Dale Cruikshank (in *Icarus*), Audouin Dollfus (in *L'Astronomie*), and Andy Young (in *Sky & Telescope*), all of whom would become good friends and colleagues. Inevitably, however, there were naysayers—not, fortunately, among the astronomers. One was a professional Polonius (I mean historian), who in a review in *Isis* gave the book a failing grade, on the grounds that I,

> a psychiatrist by profession [not quite; I was still a resident] and amateur astronomer by avocation, seems unaware of the historical literature. [If only] Sheehan had developed an analytical model of perception and the ways it functioned in the cognitive, social, cultural, psychological, technical, and political context that included both observers and observations as well as audiences that made use of the data. . . . But the book was written in isolation: neither the acknowledgements nor the footnotes indicate that Sheehan ever sought to enter the scholarly dialogue. Like many of the amateurs of whom he writes, Sheehan tried to go it alone.

What this individual failed to appreciate was that if I had tried to develop my ideas as a dissertation in some formal interdisciplinary program, like any of those I had considered at Chicago, it would probably never have been written at all. I could have spent years preparing myself for it, sharpened my pencils again and again, and never gotten around to putting down a first word. The fact that *Planets and Perception* existed at all was very much a case of a fool rushing in where angels feared to tread. I had "tried to go it alone," and for me, at least, no other way would have been possible. What the writer of that rather nasty review failed to appreciate was, moreover, that until I wrote what I did, I had no passport to join what Dante called the "eternal conversation," to which I was welcomed by those who were favorably disposed to what I was trying to do. As Cruikshank (now a close friend) wrote to me long afterward when I reminded him of his encouragement when I needed it most:

> I think all of us entered this field [planetary science], which is populated by some really smart people, feeling a sense of insecurity that we could actually succeed. Adding to a basic insecurity that is endemic to us all, many of us carry scars from sour events or hypercritical people in our college and university, perhaps most often from our thesis advisors. Consequently, even a small word of encouragement from an established person in the field is not only welcomed, but remembered and cherished for years to come. It seems to me that this makes it all the more important to offer encouragement and compliments when deserved, and I would ask all my friends to keep this in mind when working with young early career scientists and writers, whether it be in personal contacts or in other contexts, such as in the review of papers and books for publication.
>
> And in particular, Bill, I had no idea how important my remarks in the review of *Planets and Perception* would become for you. I'm touched.[4]

With *Planets and Perception*, I established what has been the main theme of my work in history of science ever since: the human side, the fallibility of the human mind, and the fact that, for all the scientific method's flaws, and "stodgy and grumpy as it may seem,"[5] it is in the end the only way humans have found to get past illusions, deceptions, pseudoscience, and lies to im-

---

4. Dale P. Cruikshank, pers. comm., August 7, 2021.

5. Carl Sagan, *The Demon-Haunted World: Science as a Candle in the Dark* (New York: Ballentine, 1993), 22.

proved understanding, and though it may not often conform to what the heart desires, it is in the end tethered to reality and closer than any other method to the truth (but always the truth accompanied by error bars). In the end, as Carl Sagan has said, "Science alone teaches us about the deepest issues of origins, natures, and fates—of our species, of life, of our planet, of the Universe. . . . All of us feel goosebumps when we approach these grand questions. In the long run, the greatest gift of science may be in teaching us, in ways no other human endeavor has been able, something about our cosmic context, about where, when, and who we are."[6]

Coming at knowledge from both ends—from the brain on one end, and the Earth, solar system, and universe on the other—I have felt privileged to have contributed, in a small way, to this grand human/suprahuman quest. I have done so while professionally (as a psychiatrist) trying to contribute my mite to the mitigation of human misery. Though I don't feel particularly successful in that endeavor—coming up against the intransigence of human nature—I have also been able, with the forbearance and support of my family, to continue to pursue writing about astronomy and the history of astronomy in books such as *Worlds in the Sky* (Arizona, 1992); *The Immortal Fire Within: The Life and Times of Edward Emerson Barnard* (Cambridge University Press, 1995); *The Planet Mars* (University of Arizona Press, 1996); *In Search of Planet Vulcan*, with Richard Baum (Plenum, 1997); *Epic Moon*, with Thomas A. Dobbins (Willmann-Bell, 2001); and—more recently— *Galactic Encounters*, with Christopher J. Conselice (Springer, 2015); *Discovering Pluto*, with Dale P. Cruikshank (Arizona, 2018); and *Discovering Mars*, with Jim Bell (Arizona, 2021).[7]

---

6. Sagan, *Demon-Haunted World*, 38.

7. Sheehan, *Worlds in the Sky: Planetary Discovery from Earliest Times Through Voyager and Magellan* (Tucson: University of Arizona Press, 1992); Sheehan, *The Immortal Fire Within: The Life and Times of Edward Emerson Barnard* (Cambridge: Cambridge University Press, 1995); Sheehan, *The Planet Mars: A History of Observation and Discovery* (Tucson: University of Arizona Press, 1996); Baum and Sheehan, *In Search of Planet Vulcan: The Ghost in Newton's Clockwork Universe* (New York: Plenum, 1997); Sheehan and Dobbins, *Epic Moon: A History of Lunar Exploration in the Age of the Telescope* (Richmond, Va.: Willmann-Bell, 2001); Sheehan and Conselice, *Galactic Encounters: Our Majestic and Evolving Star-System, from the Big Bang to Time's End* (New York: Springer, 2015); Sheehan and Cruikshank, *Discovering Pluto: Exploration at the Edge of the Solar System* (Tucson: University of Arizona Press, 2018); Sheehan and Bell, *Discovering Mars: A History of Observation and Exploration of the Red Planet* (Tucson: University of Arizona Press, 2021).

Along the way I have had many great mentors, colleagues, and countless friends, all around the world, for it is true, as Audouin Dollfus once remarked to me, because of astronomy's universality—the sky is common to all of us—as soon as we set foot on an airport tarmac, we are already assured of having fast friends for life. I could name many mentors and colleagues. Many of them are named above, as coauthors of books we wrote together. In addition, one person stands out for special recognition: Donald E. Osterbrock, known to his many friends simply as Don, astronomer, Lick Observatory director, and astronomical historian, who effectively adopted me as his protégé, and without whose mentoring I could never have finished my biography of E. E. Barnard, which I still consider probably the best thing I've ever done. (He also arranged for me to attempt to repeat, in a spirit of emulation, Barnard's observations of the planets with the 36-inch refractor at Lick Observatory, which was a revelation to someone as interested as I had been in the art of visual planetary observing.)

Above all, I count myself lucky to have been not only a student and historian of planetary astronomy, but an observer with deep roots in amateur astronomy. As such, I have knowingly had photons from distant worlds impact and travel directly to my retina (that extension of the brain, connected to the brain proper by the optic nerve). This has given me a more intimate connection with the universe than many historians of astronomy seem to have.

It All Began with Venus

KLAUS BRASCH

We are the cosmos made conscious and life is the means by which the universe understands itself.

—Brian Cox, *Wonders of the Universe (Messengers)*,
BBC TV documentary, 2011

One of my most vivid childhood memories dates back to early January 1945 in the twilight of World War II. I had just passed my fourth birthday, and the Soviet Red Army was on the doorsteps of Danzig (now Gdansk), my mother's birth city on the Baltic Sea. Allied air raids were relentless, along with sleepless nights huddled in the cellars of buildings still standing. The terrifying howls of air raid sirens haunt me to this day. Fortunately, my grandfather managed to secure passage for my mother and me on one of the last ships out before the historic city was destroyed under heavy bombardment, and my grandparents perished. As we hurried toward the port, on a frigid but crystal clear night, I spotted a dazzling bright object in the sky. When I asked what that was, Grandfather replied, "Das ist die Venus" (That's Venus). All I heard was "Nuss," or "nut" in German, and according to my mother, for long thereafter I excitedly called out "Das ist die Nuss" whenever I spotted a bright star in the sky.

Miraculously mother and I survived unscathed a harrowing seven-day trip from Danzig to Kiel in western Germany, still under Nazi control. That was mainly because the converted ocean liner *Wilhelm Gustloff* kept close to the Baltic coastline under a thick blanket of fog, thereby eluding Allied air and sea forces. The ship was subsequently sunk on its final trip, with thousands of casualties. From Kiel we gradually made our way by rail to Weimar, where my father's family owned a still largely intact house. At this point we had no idea where Father was nor whether he was still alive. Dad served in a maintenance squad of the Luftwaffe the entire war because he was an electrical engineer, and unbeknownst to us, he had surrendered to the French before Germany was liberated by the Allies. Mother and I were jubilant when American forces entered Weimar and commandeered our house but let us stay there. I vividly recall Sherman tanks rolling along our street and GIs handing out cigarettes to eager adults and chewing gum to kids. Panic struck, however, a few weeks later when U.S. troops withdrew and the Red Army moved in after the Yalta Conference, where Winston Churchill, Franklin Delano Roosevelt, and Joseph Stalin divided Germany into four occupation zones. Having no place else to go, we stayed put and considered ourselves lucky that the Soviet regional commander also requisitioned our house and let us stay, which provided us a measure of protection. Two years later we managed to escape the Iron Curtain and rejoin my father in Strasbourg, France. Since he spoke fluent French, was an experienced engineer, and was never a member of the Nazi party, Dad was quickly released from POW camp and offered a job with a radio company, in the charming alpine town of Albertville, where I first attended school and quickly learned French. Someone later told me that when it came to learning languages, young children's brains are like sponges that "absorb" them instantly.

A year later we moved to Rome, Italy, and lived with my father's relatives in a large hilltop villa. I was now in second grade and had to learn Italian. The school run by nuns was rigidly Catholic and focused on the basics of beautiful penmanship, arithmetic, and of course catechism and countless saints. My most interesting teacher in third grade was one of the few laypersons at the school, who taught us all about the glories of ancient Rome, evidence of which surrounded us everywhere. In addition to its rich history and splendid art and culture, the Eternal City also boasted a wonderful zoo and natural history museum, both of which I frequented. My favorite was a gallery depicting the solar system, dinosaurs, early Earth history, as well as a large iron meteorite. The notion of a time so far before ancient Egypt, Greece, and Rome intrigued me, even though I could not fully grasp that then. One night, Dad took me to a rooftop balcony to observe the Moon with an old brass telescope. That was a pivotal moment in my preteens; seeing that rugged landscape spawned the realization that we are not alone in space but circled by another world vastly different from ours yet so near one could almost touch it. My mother (née Fauth) jokingly maintained that my fascination with the Moon was predestined, since a relative of hers was the famed if controversial twentieth-century selenographer Philipp J. Fauth.

With little opportunity for work in postwar Italy, my parents, like thousands of other European families, looked abroad, and three years after my sister Patrizia was born, we emigrated to Canada in 1953, in what was truly the New World for us, new language, new culture, new challenges. We took it all in stride, though, grateful for new beginnings far from stagnating postwar Europe. Like most immigrant families in Toronto, my parents took whatever work was available, and I delivered newspapers before school early mornings. As money was always an issue, that was my allowance at the time. I did not mind those days getting up before dawn and being enthralled on clear nights by the stars overhead, still readily visible then, even in a large city like Toronto. I did well in school, quickly mastered English, and was again drawn to science and nature, with my parents' encouragement and access to the wonderful Royal Ontario Museum, now a world-class institution. Summer weekends, the family would go for picnics in one of the many city parks and to the Humber River valley to collect fossils dating back to the Late Paleozoic, 450 million years ago. That's also when the seeds of biology became planted in my consciousness. My parents gave me a small Japanese-made microscope, complete with glass slides and a how-to book to examine everything, from hair, onion skin, and pond water to small insects and blood cells. The microworld was as captivating to me as the Moon had been through a telescope.

A particularly formative year for me was 1957, the beginning of high school, with real science classes and, almost like an omen, magnificent comet Arend-Roland gracing the then still dark Toronto skies, which really perked my interest in astronomy. For a brief moment I actually thought I might have discovered it, only to read otherwise in the next day's newspaper. Along with that, Sputnik 1 was launched, much to everyone's shock, since we all thought the Americans would do that first. My father had just built a short-wave radio, letting us listen to its haunting beep-beeps every time it passed overhead. The mere realization that humanity had now entered "outer space," as we called it then, led to flights of fancy about reports of "flying saucers," which seemed to pop up regularly in newspapers and radio news. A buddy and I even contrived to build a Cartwright Saucer Detector we read about in a pulp magazine, as well as in books by George Adamski, one of the most successful UFO snake oil sellers of all time. Fortunately, my parents' skepticism and Martin Gardner's sobering book *Fads and Fallacies in the Name of Science* soon put me back on track. Still, there was always the classic 1953 movie rendition of H. G. Wells's *War of the Worlds*, as well as such masters of science fiction as Isaac Asimov and Arthur C. Clarke, to help permanently fix the notion in me that intelligent beings must be out there, and we might contact them someday.

The International Geophysical Year (IGY) was in progress; I was now old enough to join the Royal Astronomical Society of Canada (RASC); and, with my father's help, I built my first decent telescope, a 3-inch refractor, with war-surplus optics, pipe mount, and homemade accessories (figure 7.1). It showed me Saturn's glorious rings, Jupiter's captivating moons, sunspots, exquisite lunar details, and the splendid Orion Nebula. Eventually, though, after seeing many seductive advertisements in *Sky & Telescope*, I succumbed to a case of "aperture fever." Since cost was always a limiting factor, I could only afford the optics for an 8-inch f/7 Newtonian reflector made by legendary mirror maker Alika K. Herring, at Cave Astrola in Anaheim, California. Again with Dad's help and some spare parts, I built what was then considered a large amateur instrument, which really broadened my observational horizons.

The Toronto Center of the RASC, by far the largest in the country, was a rather stodgy, formal group of mainly older men, who seemed more dedicated to perfunctory things like the president, secretary, and treasurer's reports each meeting than what a novice like me was there for. I soon discovered, thankfully, that this was not totally the case, and that the membership fell into three informal groups: armchair astronomers, observers, and telescope makers. Naturally, I gravitated toward the observers and first met Terence Dickinson, who later became a close friend and a leading astronomy author and popularizer. Years later, his now classic *Night Watch* would become one of the best-selling amateur astronomy guide books ever.

We moved to Montreal in 1958. My father's company needed a French-speaking engineer there, and I was enrolled in what proved to be one of the best high schools in the city. I quickly joined the RASC Montreal Centre, one of the most active in North America. Unusual for those times, the center boasted a first-rate observatory of its own, complete with a fine 6-inch refractor, well-stocked library, meeting hall, and telescope-making workshop. Located on the McGill University campus, the observatory building had been an experimental radar facility during World War II and was subsequently repurposed for astronomy by club members. Led by a remarkable

FIGURE 7.1 My father and friend with our 3-inch homemade refractor at the July 20, 1963, total solar eclipse, eastern Quebec. Courtesy of Klaus Brasch.

woman, Isabel Williamson, the center issued a monthly newsletter (*Skyward*) and held frequent public star parties. With mentorship by the center's elders, mostly people with day jobs but a passion for astronomy, newly minted members were soon exposed to activities like identifying lunar features, the Messier Club, lunar occultation timings, nova searches, observing and sketching the planets, daily sunspot counts, and, most helpful, encouragement to give public talks at club meetings. All this helped instill self-confidence in shy teens like me and appreciation of the joy and intellectual stimulation astronomy provided.

Spurred on by *Sky & Telescope*, everybody's astronomy bible, anxiously awaited each month and read from cover to cover, as well as many engaging "Guide to" books by the prolific and accessible British popularizer of astronomy Patrick Moore, it was deemed possible for amateurs like me to make valid contributions to science, even with modest telescopes and largely visually. Many key basics about the Moon and planets were still unknown. Among these were the rotation periods of Mercury and Venus, the origin of lunar craters, and the atmospheric subtleties of Jupiter and Saturn. Not until Gene Shoemaker showed definitively that lunar craters were the result of impact events, not volcanism, did that point of contention resolve. Likewise, surface conditions on Venus fell into three theoretical if contradictory scenarios: steaming jungle, water world, or parched desert. Although the artificial canal theory was largely abandoned, it still seemed plausible that seasonal darkening of Martian albedo features might indicate some form of vegetation.

Consequently, in 1960, our group of young enthusiasts, with aspirations of becoming astronomers, began monitoring the Moon and bright planets telescopically and in coordinated manner, using telescopes of 3 to 8 inches in aperture and Wratten series color filters. This included looking for the ashen light, changes in Venusian cloud features through deep violet filters, and tracking its apparent phase anomaly (the so-called Schröter effect). We also undertook extensive central meridian timings and strip sketches of Jovian cloud features and the Red Spot. This was part of a long-standing British Astronomical Association (BAA) program to monitor differential drift rates in the planet's atmosphere at various latitudes.

The ashen light on Venus is reminiscent of earthshine on the very new and old Moon, somewhat doubtfully noted by Giovanni Riccioli in 1643 and subsequently by others, though never consistently. It has been variously attributed to optical illusions, Venusian lightning, thermal radiation, oxygen emissions, and reactions to solar X-flares. While none of these are entirely convincing, recent results from the Parker Solar Probe suggest that CO_2 is split into its constituent ions by intense UV radiation on the day side of Venus and could recombine into O_2 on the night side to emit faint green light potentially detectable by the eye.[1]

---

1. Brian E. Wood et al., "Parker Solar Probe Imaging of the Night Side of Venus," *Geophysical Research Letters*, February 9, 2021, https://doi.org/10.1029/2021GL096302.

Johann Schröter first reported the Venus phase anomaly in 1793, by noting that the observed dichotomy (or half phase) differed from the predicted by six to eight days, early for eastern and late for western elongations. These values were in line with those we obtained in Montreal during the 1961 and 1962 western elongations. As with the ashen light, several explanations have been advanced for this curious effect, ranging from atmospheric twilight scatter to optical illusions and differences in orbital eccentricities between Earth and Venus, among other considerations. A complete explanation is likely to depend on fuller understanding of the physical properties of the lower Venus atmosphere, or mesosphere.

Although the 1960–61 opposition of Mars was not particularly favorable, our team dedicated much effort to observing and sketching its albedo features and polar regions through color filters. Thanks to clear weather and good seeing conditions, we obtained enough decent observations for member Geoffrey Gaherty Jr. to construct a quality albedo map of the planet for his college geography class project (figure 7.2). Note that we all sketched linear canal-like features, unsure about what they were. Though few of us thought them artificial, we still hoped against hope that they might be natural water channels, as Earl C. Slipher of Lowell Observatory, the great Mars authority at the time, and other professionals maintained. We also presented

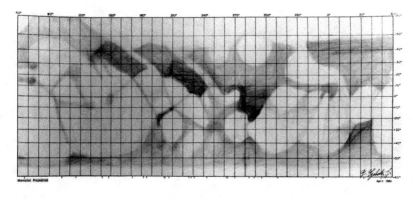

FIGURE 7.2 Map of Martian albedo features based on some fifty observations made by members of Montreal Centre during the 1960–61 opposition, as compiled by Geoffrey Gaherty Jr. Courtesy of RASC archives.

reports of our work at annual conventions of the RASC and the Association of Lunar and Planetary Observers (ALPO).

In addition to copious visual observations of the Moon and planets, I ventured into astrophotography. It was still a challenging undertaking in those days. Since any modern camera was way out of my price range, I adapted a toy Bakelite 35 mm camera and leaf shutter, with wax paper as the focusing screen. While the very grainy results left much to be desired (figure 7.3), they were good enough to hook me on astro-imaging to this day.

No doubt the most important scientific projects we participated in were systematic studies of meteor showers under the direction of prominent astronomer Peter M. Millman. As a student at Harvard University, Millman was prompted by none other than Harlow Shapley to study meteors and their spectra, a novel idea at the time, and something he pursued at the National Research Council of Canada (NRCC) his entire career. To help with this massive project, Millman recruited amateur groups across the country to collect data on the frequency, direction, magnitude, and color of meteors during major showers, information very much needed as a prelude to manned spaceflight, since it was still unclear how much of a hazard they might present. We participated eagerly and learned much about research and teamwork.

Similar to Operation Moonwatch during the IGY in 1957–58, when amateur teams tracked artificial satellites, we organized into groups of six or more observers and a time keeper/recorder. After setting up comfortable

FIGURE 7.3 *Top*, modified Bakelite camera the author used to get, *right*, his first photograph of the Moon. At the time this photo was taken (around 1960), this was a rather creditable result. Courtesy of Klaus Brasch.

outdoor recliners at a dark location, each observer, armed with red lights and specially prepared star charts, would watch a selected sector of the sky. We called out "time" whenever a meteor was spotted, and plotted its direction and magnitude on our charts. The data then went to Millman in Ottawa for compilation and analysis. These activities were fun and instructive, but they also required considerable stamina, against mosquitoes in warm weather and the oft bitter cold of clear Canadian winter nights, though foul-smelling bug repellant in summer and copious cups of hot chocolate in winter helped considerably.

A pivotal event for me as a student was meeting one of the true giants of astronomy, Cecilia Payne-Gaposchkin. Each year the Montreal Centre used a small endowment to invite a famed scientist for a special speaking engagement and visit. After a riveting lecture on stellar chemistry and evolution, Payne joined us for an evening at our observatory and regaled us with stories of her student days in England and then as the first female astronomy professor at Harvard University. Later, chatting with her on the observatory's deck on that lovely fall night, she first made me aware that all the stars we see by naked eye are actually relatively close neighbors standing out against the distant backdrop of the Milky Way. That was a true eureka moment for me, now appreciating for the first time the imposing 3D panorama of our home galaxy. She also ardently encouraged us as students to pursue our dreams of becoming scientists. This inspiring quotation of hers tells us why: "The reward of the young scientist is the emotional thrill of being the first person in the history of the world to see something or to understand something. Nothing can compare with that experience. The reward of the old scientist is the sense of having seen a vague sketch grow into a masterly landscape."[2]

I enrolled at Sir George Williams University (now Concordia) in Montreal in 1959 and majored in biology, because mathematics was not my forte, and I had thoughts of a future biomedical career. Shortly after, my dad was diagnosed with a pituitary tumor and a grim prognosis. Fearing the worst, I began looking for work to support my mother and young sister, and I failed my classes miserably. Thankfully, my father's life was saved by a then highly experimental procedure known as cobalt external-radiation therapy, at the Montreal Neurological Institute-Hospital. With that reprieve, I approached

---

2. Cecilia H. Payne-Gaposchkin, "Fifty Years of Novae," *Astronomical Journal* 82, no. 9 (1977): 665–73.

my second year of college with new determination and also met the love of my life and future wife, Margaret Schoning, in an introductory biology class.

Many active Montreal Centre members also joined the ALPO, then under the capable direction of Walter H. Haas. He had founded the organization in 1947 and encouraged amateurs around the world to monitor the Moon and planets and submit observations for reduction and reporting in the *Strolling Astronomer*. Many future astronomers and scientists, myself included, benefited greatly from Walter's gentle and ever astute mentoring. I first met him at the ALPO convention in Detroit, Michigan, in 1961 (figure 7.4) and again in 1962 at the tenth convention in Montreal, the first outside the United States. Much to my surprise, and because I had submitted many Mars observations to Walter, he offered me the position of ALPO Mars recorder, which I accepted gladly and held until moving to Ottawa for graduate studies in 1965. My successor would be Charles (Chick) Capen, then resident astronomer at the Jet Propulsion Laboratory's Table Mountain Observatory, which helped provide support to the Mariner missions to Mars, and later a staff astronomer on the long-running International Planetary Patrol Program at Lowell Observatory.[3]

---

3. The International Planetary Patrol became operational in 1969 and involved a collaborative effort among observing sites that would allow the planets to be kept under as continuous photographic surveillance as possible. Initially, the Lowell Observatory; the Magdalena Peak Observatory of New Mexico State University; Mauna Kea in Hawai'i; Mount Stromlo Observatory near Canberra, Australia; the Republic Observatory in Johannesburg, South Africa; and Cerro Tololo in Chile were involved. Magdalena Peak participated for only one year, but in 1970 the Astrophysical Observatory at Kodaikanal in southern India joined, and Perth Observatory in Western Australia substituted for the Mount Stromlo station. Mauna Kea, Cerro Tololo, and Perth were equipped with new 24-inch Boller and Chivens telescopes; and new optics, to make them perform identically to the new telescopes, were installed in existing ones at Lowell and Kodaikanal—including the historic 24-inch Clark refractor on Mars Hill. Initial emphasis in the program was on Mars and Jupiter, though later UV images of Venus and a few images of Saturn were added. In the first three years, a total of fifty-six thousand imaging sequences were obtained, with each sequence containing fourteen images. The full patrol effort continued only until 1976, and scaled-back versions until 1990, when the program was canceled. Unfortunately, it did not achieve results commensurate with the effort—largely because the images were registered on Kodachrome film, whose graininess did not allow recording the finest planetary details. Ironically, the program ended just as CCD imaging was becoming available. Brasch, who attempted to digitally scan some of the images, has summarized the IPP program from promise to disappointment in "The Dawn of Global Planet Watches," *Sky & Telescope* 131, no. 1 (2016): 52–54.

FIGURE 7.4 An intense moment at the 1961 ALPO convention in Detroit. *Left to right,* Walter Haas, Klaus Brasch, Geoff Gaherty, and Clark Chapman. Courtesy of Klaus Brasch.

My astronomical activities went into prolonged hibernation during my first year in grad school at Carleton University, as I was now fully occupied with research, coursework, and as a teaching assistant. Happily, Margaret and I married the year after, and she too was accepted at Carleton to pursue a master's degree. Our graduate school years were some of the most exciting and memorable of our lives. Carleton had recruited some top-tier scientists from the NRCC, whose philosophy was totally hands-on instruction, still a novel concept at the time. Ironically, much of the advanced math that had discouraged me from majoring in astronomy as an undergraduate now proved essential for the theory and practice of electron microscopy, one of my primary research tools. The demands put on us as students were high, but the rewards even higher.

After obtaining both master's and PhD degrees in cell biology, and after two years of postdoctoral work, I started looking for a job. Although offered a position in the State University of New York system, I opted for Queen's University in Kingston, Ontario, in 1973, where I also found some time again for astronomy. While I chose a biomedical career, astronomy has always provided an overarching perspective for me as scientist, researcher, teacher, and human being. One of my favorite Carl Sagan quotations, from his book *A Pale Blue Dot*, says it all: "It has been said that astronomy is a humbling and character-building experience. There is perhaps no better demonstration of the folly of human conceits than this distant image of our tiny world. To me, it underscores our responsibility to deal more kindly with one another and to preserve and cherish the pale blue dot, the only home we've ever known."[4]

As a young biology professor at Queen's, I met and befriended astronomer Alan Bridle, who had been granted time on the 46-meter radio telescope at nearby Algonquin Radio Observatory for one of the very first SETI searches. Like me, Alan was fascinated with the larger question, Are we alone in the cosmos?, and the newly established SETI efforts were the impetus we needed to team up for an interdisciplinary course, Planets and Life. Despite being teased relentlessly by our colleagues about "little green men," our course quickly became one of the most popular at the university, and it showed me that most people are intrigued by the possibility that life may be widespread in the universe. In light of that, I became interested in the emerging science of astrobiology and questions about the origin of life on Earth and elsewhere, an interest that remains to this day.[5] Needless to say, these questions are as compelling as any that humans have ever asked, and they are at the core of many current planetary missions and the search for Earth-like exoplanets. Alan later moved to the U.S. National Radio Astronomy Observatory in Green Bank, West Virginia, for a very productive research career in, among other things, astrophysical jets now known to be associated with supermassive black holes in active galaxies.

Much changed during my tenure at Queen's. I was determined to make a mark in my research, teaching, and mentoring graduate students in cell and molecular biology, which were then (and remain) fields very much at the cusp of major advancements in genetics, cancer, and the emerging areas of recombinant DNA, with the promise of gene modification. These cutting-edge developments demanded total dedication to research, grant writing, publishing, and promoting one's work at conferences and work-

4. Carl Sagan, *Pale Blue Dot: A Vision of the Human Future in Space* (New York: Random House, 1994), 9.

5. See, for instance, Klaus Brasch, "Life in the Cosmos Revisited," *Journal of the Royal Astronomical Society of Canada* 116 (2020): 257–65.

shops. Amid all that, Margaret and I rejoiced in the birth of our beautiful daughter, Madeleine.

Later that year, we migrated to California for a sabbatical leave at the City of Hope National Medical Center in Duarte, where I was a visiting scientist in medical genetics, and my research sharply refocused toward cancer and cellular reprogramming. I also resumed an active interest in astronomy by visiting Mount Wilson, Palomar, and Griffith Observatories and acquired a truly classic telescope, a sixties vintage Celestron-10 produced by the Torrance, California, manufacturer of affordable Schmidt-Cassegrains. I upgraded to a Celestron-14 before the favorable 1988 Mars opposition, when the disk diameter reached nearly twenty-four arcseconds at closest approach to Earth, and the planet stood almost on the celestial equator (hence reasonably well placed for Northern Hemisphere observers). This was the last opposition at which films were still used almost exclusively to record planetary images, just as Earl C. Slipher had done. A few professional astronomers, such as François Colas and Jacques Lecacheux at Pic du Midi Observatory in France and Stephen Larson at the Catalina Station of the University of Arizona observatories, were beginning to use charge-coupled devices (CCDs), but they were cutting edge. For amateurs such as me, the state of the art was the newly available, red-sensitive Technical Pan 2415 film. While this film captured more detail than I had ever obtained before, I could still detect finer, more nuanced features visually and consequently decided to add this finer detail to photos taken at the same time (figure 7.5). Many of these hybrid images were subsequently published in *Astronomy* and *Sky & Telescope*. The era of superiority of the human eye-brain-hand system was, however, almost over; indeed, in 1988 when the first spectacular CCD images of Mars, with detail reminiscent of that shown in E.-M. Antoniadi's famous early twentieth-century drawings, were published in *Sky & Telescope*, it was apparent that its days were numbered.

In May 1981 I was invited to a remarkable conference at the National Museum of Natural Sciences in Ottawa, Canada. Hosted by famed geologist and paleontologist Dale A. Russell, it was the second K-TEC workshop on Cretaceous-Tertiary Extinctions and Possible Terrestrial and Extraterrestrial Causes. (What was then known as the K-T, Cretaceous-Tertiary, mass extinction is now generally referred to as the K-Pg, for Cretaceous-Paleogene; the latest date for it is sixty-six million years ago.) One possible cause of this mass extinction was suggested by the father and son team of

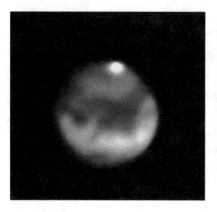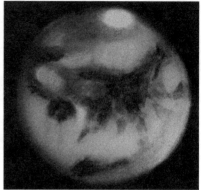

FIGURE 7.5 *Left*, photograph of Mars; *right*, a photo drawing, in which subtle details recorded visually have been added in pencil to a velour paper print of that image. Courtesy of Klaus Brasch.

Louis and Walter Alvarez. Having identified (with two colleagues) a worldwide iridium-rich clay layer at the K-Pg boundary, the Alvarezes argued that the enrichment by iridium could only have occurred via an extraterrestrial source, i.e., a colossal asteroidal or cometary impact. Some geologists and paleontologists favored an alternative hypothesis, however, that the mass extinctions occurring at this time were due to climate change associated with volcanism on an enormous scale occurring as the Neotethys Ocean plate subducted under the southern edge of Eurasia and the Indian and Eurasian tectonic plates collided (eventually causing uplift of the Himalayas). Controversy continued until the nineties, when the Chicxulub crater in the Gulf of Mexico's Yucatán Peninsula was identified. This showed that the asteroidal impact had indeed occurred and was at least partly responsible for the mass extinction in which some 80 percent of animal species were wiped from the face of the Earth (including most famously the dinosaurs). Since then, it is generally agreed that it was not only one factor but rather a one-two blow in which the impact followed by volcanism on an enormous scale combined to produce the cataclysm.

Among the many topics at the workshop, one of the most intriguing was: Had the mass extinction not occurred, would we be here today? Russell was among the first dinosaur experts to posit that many were at least in part warm blooded. This was based on two main criteria: the much higher density of blood vessels in fossilized dinosaur bone marrow compared to reptiles, and rising evidence that modern warm-blooded birds are the direct descendants of dinosaurs. Since mammals composed but a minor class of vertebrates at the time, absent the impact, would they have risen to dominance postextinction as they did, and would intelligent primates including us have evolved? Russell and artist Ron Sequin undertook a bold thought experiment, suggesting that intelligent beings might well have evolved along a line of now extinct birdlike dinosaurs. Much to everyone's delight, they then unveiled a life-size model of *Dinosauroid* (figure 7.6).

FIGURE 7.6 *Dinosauroid* sculpture by D. Russell and R. Sequin, 1982. Courtesy of Canadian Museum of Nature.

In 1990, I was offered a position in biology at California State University, San Bernardino, and tasked with enhancing the research and grant acquisition potential of the department. As well as a hub of cutting-edge biomedical science, California was a haven for astronomy, with several major observatories and telescope manufacturers. I took full advantage of that, both academically and in reawakening my passion for all things astronomical. Weather and time permitting, I joined a group of stargazing friends for weekend sorties to spectacular dark sites in the mountains and deserts, honing my astrophotography skills, first with film and then digitally.

In the late nineties, Margaret and I were contracted by Cambridge University Press and Firefly Books to translate French and German astronomy and popular science books into English. These included three outstanding works, *Great Observatories of the World* by Serge Brunier and Anne-Marie Lagrange; *New Atlas of the Moon* by Thierry Legault and S. Brunier; and *Space Probes* by Philippe Séguéla, all for Firefly Books. I also began writing articles for *Astronomy*, *Sky & Telescope*, the *Journal of the RASC*, and others. That kept me abreast of astronomical developments, for talks to local clubs, and teaching the very popular course Life in the Cosmos at the university. The latter, more than any other teaching I had done, showed me, once again, that no matter what their interests or backgrounds, most people are both awed and intimidated by the vastness of the universe and our place in it. In the words of eighteenth-century physicist and philosopher Georg Christoph Lichtenberg: "Astronomy is perhaps the science whose discoveries owe least to chance, in which human understanding appears in its whole magnitude, and through which man can best learn how small he is."[6]

More important, astronomy provided a welcome intellectual balance to my demanding research, administrative, and academic responsibilities at the university. I served first as department chair, then dean of the college of science, and later as director of campus research. In that capacity I secured the university's first commercial patent based on novel electronic technology, developed by two outstanding young physicists. This earned me the Research Advocate of the Year Award in 2004 from the U.S. Small Business Administration, and an Outstanding Service Award from the university the following year. Margaret, as director of the institution's grants office, secured a lucrative grant from the Keck Foundation (of W. M. Keck Observatory fame) for a university student research observatory, which earned her the Outstanding Staff Award. I was also invited to collaborate with researchers at the world-renowned Scripps Research Institute in La Jolla, California, to extend my work in cellular differentiation and cancer, leading to some of the most significant publications of my career.

During my now long journey as an astronomy enthusiast, teacher, and student, I was fortunate to have observed and experienced some truly memorable things; many were just minor, personal firsts, others really extraordinary and unique. Among the former I include spotting Triton, Neptune's largest moon; the Martian crater Schiaparelli; and Pluto, all with my C-14 under exceptional seeing conditions. Among the truly astonishing, I recall a massive coronal-type aurora that lit up the sky over Montreal, a Leonid meteor shower with more than one thousand per hour from the Mojave Desert, dazzling comets Hyakutake and Hale-Bopp from the same location, stunning views of Uranus and some of its major moons with the venerable 60-inch reflector at Mount Wilson Observatory, the unparalleled 1991 solar eclipse in Baja California, the 2012 transit of Venus at Lowell Observatory, and a transit of Io across Jupiter under superb seeing with the outstanding 4.3-meter Lowell Discovery Telescope.

The July 11, 1991, solar eclipse featured one of most spectacular and longest episodes of totality possible, at nearly seven minutes. Sited in La Paz,

---

6. Georg Christoph Lichtenberg, aphorism 23, Aphorisms (1765–1799), Notebook C (1772–1773), as translated by R. J. Hollingdale, in *Aphorisms* (New York: Penguin, 1990).

Mexico, the charming capital of Baja California, thousands of eclipse chasers from around the world and I enjoyed superbly clear and transparent skies for the event. Quite coincidentally, I was set up next to Dale Cruikshank (a contributor to this book) and his family. The darkening of the midday sky, dramatic drop in ambient temperature, odd behavior of birds and other animals, last blaze of sunlight in a "diamond ring" effect, and a dazzling corona at totality brought screams, cheers, and tears from the assembled multitude. Small wonder ancient cultures were terrified by such spectacles! Terrified— but also relieved, when the apparent "death" of the Sun ended with the reappearance of a sliver of light signaling the resumption of normalcy.

My appreciation of the 2012 transit of Venus was greatly enhanced thanks to participating in an international effort led by Paolo Tanga (a contributor to this book), astronomer from Observatoire de la Côte d'Azur, who visited Lowell Observatory for the event. Known as the Venus Twilight Experiment, it focused on the "aureole" visible within the halo of the Venusian atmosphere as it transited the Sun's corona. As a measure of the refractive properties of the planet's atmosphere, this provided information on the physical properties of the lower atmosphere, or mesosphere, of Venus and will likely assist with similar studies of exoplanets transiting their home stars. Tanga and colleagues modified small portable telescopes as coronagraphs for this purpose, William Burke and I provided equatorial mounts, while William Sheehan (a contributor to this book and co-author of a definitive reference work on transits of Venus) assisted with the actual observations.

The most truly overwhelming experiences for me, however, were seeing and photographing the southern Milky Way and Magellanic Clouds in 1999, when my old friend Terence Dickinson and I were generously granted ten nights' observing time at Australia's preeminent optical observatory at Siding Spring, and later from San Pedro de Atacama in Chile. Our images were then used in Terry's many popular astronomy books and articles and in my teaching efforts. Those who have been to the Southern Hemisphere, and Atacama in particular, can vouch for the majesty of the Milky Way in all its splendor, which, when seen for the first time, can be a truly overwhelming and emotional experience. When directly overhead under those splendid and transparent southern skies, the core of the Milky Way actually casts a shadow (figure 7.7). In addition to unparalleled astronomical vistas, northern Chile also offers some of the most imposing geological terrain on the planet, including snowcapped Andean volcanoes and the Chajnantor Plateau, site of the Atacama Large Millimeter Array (ALMA).

FIGURE 7.7 Milky Way photographed in 1999 from Siding Spring Observatory is among the author's best film-era astronomical images. Courtesy of Klaus Brasch.

In 2006, Margaret and I retired to Flagstaff, Arizona, "The world's first international dark sky city." We supported Lowell Observatory for many years, and I now had the pleasure to volunteer in its popular public program (figure 7.8). With occasional access to its storied 24-inch refractor, I obtained my first digital images of the Moon. One has a deep sense of history using this superb Alvan Clark and Sons instrument, which, along with Lowell Observatory's rich archival collections, has provided ample fodder for articles in astronomy magazines and journals. Thanks to its legendary dark skies, Flagstaff is also peppered with amateur astronomers, many with backyard observatories, who, through the Coconino Astronomical Society, participate in special events at Lowell and the annual Flagstaff Festival of Science and Star Party. These events attract thousands of local and external visitors, offering a unique opportunity to educate young and old alike about science, art, and nature.

Without doubt, the highlight of my lifelong passion for astronomy was the naming of asteroid 25226-Brasch = 1998 TP30, with the following citation: "Klaus Brasch began volunteering at Lowell Observatory in 2008 as a portable telescope operator in our evening outreach programs. He is also an astrophotographer and writer; many of his images as well as Lowell-related articles have been published in *Sky & Telescope* and *Astronomy* magazine, among others."

FIGURE 7.8 The author and Lowell Observatory historian Kevin Schindler preparing for an observing session with the historic 24-inch Clark refractor. Photo by Richard Edmonds. Courtesy of Klaus Brasch.

I am occasionally asked if I ever regret not choosing astronomy as a professional career. On the contrary, my blended interests in both the micro and macro universes have helped me to fully appreciate not only the grandeur of nature but also our place in it. Looking back on my life and career in science and academia, now more than ever, I see humanity, our planet, and its future from a cosmic perspective, and at a time in history when answers to such long-standing questions as What is life?, Is it universal?, and Are we alone?, may actually be answered. But perhaps more pressing, can we save our planet from such existential threats as wars, overpopulation, pandemics, inexorable climate change, and environmental catastrophe? I remain optimistic that we can, but only if we unite and finally see spaceship Earth—the "pale blue dot," recalling Carl Sagan's famous phrase—as the only home we have, and most likely the only one we will ever have.

There's No Such Thing as Now

WILLIAM LEATHERBARROW

And later times thinges more unknowne shall show.
Why then should witlesse man so much misweene,
That nothing is, but that which he hath seene?
What if within the moones fayre shining spheare,
What if in every other starre unseene
Of other worldes he happily should heare?

—Edmund Spenser, *The Faerie Queene*

Memory is an act of both recovery and betrayal. The reality of those events it recovers is often betrayed by the intervening years of imaginative re-creation.

For example, in my memory I have been an amateur astronomer all my life. I remember seeing the first episode of Patrick Moore's *The Sky at Night* on BBC television on April 26, 1957. In my mind's eye I can clearly see Arend-Roland, the great comet that graced our skies in that same month. And I remember the launch of Sputnik 1, which ushered in the space age in October of that same momentous year.

In reality only the last of those memories is real. As a ten-year-old raised on comics, I was fascinated by anything to do with spaceflight. I was much less aware at that age of the science of astronomy itself, and I was almost oblivious to what little could be seen in the orange, light-polluted urban skies above my home in Liverpool.

Sputnik changed everything. I followed avidly all subsequent developments in the evolving space race: the first dog in space, the first crewed flights, the earliest robotic missions to the Moon and planets, the first images of the Moon's averted hemisphere, and the mounting excitement of humankind's attempts to land a man on another world. Those events populated my years of formal education. They started as I was about to enter the Liverpool Institute High School for Boys—the same school attended by Paul McCartney and George Harrison, who went on to achieve a certain level of success in another field—and ended with the first Moon landings, which occurred when I was already a graduate, married with a child, and recently embarked on the adventure of adult life.

But the years at school had seen my interests evolve beyond a romantic enthusiasm for space travel to other worlds and into a more considered fascination with the night sky. Several factors contributed to that shift, but paramount among them was the essentially paradoxical nature of astronomy itself. At the same time, it is both the most accessible and the most intangibly elusive of all the sciences. Accessible because it is, quite literally, just above our heads, and it requires no equipment other than our own eyes to make an experimental start. Later on we will no doubt look to acquire binoculars and a telescope, but many wonders of the universe reveal themselves immediately to the unaided eye. Yet, it does not take long to sense just how difficult it is to grasp with one's reason the sights so readily available to the eye. I can still recall the overwhelming sense of awe that accompanied my first awareness of the scale of the visible universe. In particular I remember a schoolmaster telling me that if the Sun were reduced to the size of a period at the end of a sentence in a book, then on that scale, the nearest other star would be another period some 5 miles (8 km) distant. I am still not sure how accurate that analogy was, but at that instant, I first truly understood the meaning of the word *space*. A similarly jaw-dropping moment came with the realization that when we contemplate the vault of stars above our heads, the combination of a constant speed of light and different stellar distances means that we are seeing each of those stars at a different moment in time. There truly was no such thing as a universal *now*.

How far removed such intellectual enormities were from the gray, drab limits of life in post–World War II Britain, where food rationing did not end until 1954. What possibilities for escape they offered to someone raised in a working-class home in a rundown, once-great seaport that had yet to be rescued from depression and re-energized by the explosion of its popular culture in the sixties. I went on to find my career path via the humanities, but throughout my life astronomy has provided me with an absolute measure alongside which the scale of human achievement must be judged. But in this chapter, I seek to go beyond the autobiographical and use the personal odyssey of an individual amateur astronomer to shed light on both the social context within which amateur astronomy developed in postwar Britain and the fertile ground where what later became known as citizen science would be seeded. The huge enlargement of imaginative horizons unleashed by humanity's venture into space contributed significantly to the growth of scientific literacy in Britain, in those who went on to follow careers in science and among those who pursued different paths through life but who participated in the astronomical renaissance as amateurs.

In this chapter I trace the major scientific, commercial, and social factors that shaped the trajectory of amateur astronomy in Britain from the late fifties to the present day. These include such things as how the unique accessibility of astronomy as a science eventually allowed mass participation and opportunities for amateur-professional collaboration. I look at how those opportunities have evolved and how such evolution has been facilitated by increasingly sophisticated techniques available to the amateur. Moreover, the development of amateur astronomy over the period in question coincided with an age of rapid and unprecedented growth in our understanding of the universe around us. This is particularly so for solar system studies, where the ever-accelerating robotic exploration of our celestial backyard brought about a process of enlarging our horizons, a process similar to that described by Spenser in the epigraph to this chapter. Watching hitherto mysterious worlds gradually shed at least some of their mysteries was intoxicating and that almost certainly accounted for my own overriding interest in lunar and planetary astronomy.

It is easy to overlook from the vantage point of today just how little we knew about our neighboring worlds at the start of the space age. For example, many still held to the view that the craters of the Moon were volcanic, rather than of impact origin; for some the suspicion still remained that Mars might indeed be "the abode of life," albeit in the form of vegetation rather than Percival Lowell's intelligent canal-builders; and Venus was a planet of total mystery—perhaps a water world, or perhaps even covered in dense primeval forest.[1] Both planets held the possibility that they too might harbor comparatively advanced life forms. Such ideas might sound absurd to

---

1. A reference to the title of Percival Lowell's third book on Mars, *Mars as the Abode of Life* (New York: MacMillan, 1908).

us now, but at that time even some of the most fundamental physical properties of our neighboring worlds were still unknown. Did Mercury have a
captured rotation, with one face permanently exposed to the scorching heat
of the Sun? What was the rotation period of Venus, whose solid surface was
permanently hidden by its thick blanket of clouds? What was the Great Red
Spot on Jupiter and what went on beneath the ever-churning visible "surface"
of that planet? What were the rings of Saturn? Of the worlds beyond Saturn,
hardly anything was known.

Such mysteries, along with the excitement of the nascent space age, stimulated the need to see for oneself, and this meant getting a telescope. In
postwar Britain that was easier said than done. The UK had few commercial
sources and no real tradition of amateur telescope making. Occasionally, an
issue of the American magazine *Sky & Telescope* would find its way into my
hands, replete with reports of telescope-making conventions and seductive
advertisements for stunningly beautiful, but hopelessly unattainable, Unitron refractors. The few traditional British suppliers, such as Broadhurst
Clarkson, Brunnings, Henry Irving, and Charles Frank, could offer brass
3-inch refractors and even some 6-inch or 8-inch Newtonians, but at what
seemed like impossible prices. Today in the UK you can buy a fine Chinese-
made 6-inch Dobsonian for just over £250 (~$342). I have before me a 1963
advertisement for a similar telescope retailing then at around £80 (~$110)—a
sum equivalent to about £1,400 (~$1,916) today.

As a result, my first telescopes were the usual cardboard tube and specta-
cle lens affairs, which were unsatisfactory, and an early attempt at grinding
a mirror myself was even less successful. A breakthrough came with a do-
it-yourself kit for a 4-inch reflector, sold at a reasonable price by W. Ottway,
which came with a typewritten guide to using the completed item, written
by a then little-known Patrick Moore. This telescope, although very limited
in both size and quality, provided me with my first real glimpses of the Moon
and planets.

Things improved toward the midsixties with the advent of new players in
the telescope market. Most effective was Dudley Fuller, a likable and rogu-
ish musician turned telescope maker, who with his enterprise Fullerscopes
transformed the availability of affordable astronomical equipment in the UK.
With parts bought from Dudley and a mirror hand figured by the renowned
optician Henry Wildey, I was able to obtain a fine 8.5-inch Newtonian that
allowed me to make detailed observations of the Moon and planets and set
me on the path of being a serious amateur astronomer.

I would not have been able to tread that path successfully without the help of others serving as mentors and guides, and many—wittingly or unwittingly—helped me so much in that regard. I am grateful to all who proffered advice to a young novice, but three individuals in particular were lasting role models and provided constant inspiration, support, and a firm hand when needed. All are gone now, but my debt to Patrick Moore, Harold Hill, and Richard Baum endures to this day. It could be argued that the primary role of a mentor is to provide a combination of inspiration, criticism, and perspective. My three mentors certainly did that, but they did so with different emphases.

Patrick Moore was an inspiration not just to me, but to several generations of astronomers, professional and amateur, throughout the world. His popular books, starting with *Guide to the Moon* in 1955, conveyed an irresistibly infectious enthusiasm for the subject, and his clarity of style made the most obscure concepts accessible to the nonspecialist reader. His long-running BBC television series *The Sky at Night*, which started in the same year as the space age, reinforced the impact of his writings and perhaps did more than anything else to bring the science of astronomy to the attention of the British public. Until the advent of that program, there had been little coverage of astronomy, as opposed to space flight, in the popular media. Some newspapers carried monthly columns on what could be seen in the sky, but there were no magazines devoted to the subject. American amateurs had *Sky & Telescope*, but regular access to that fine magazine was unreliable in the UK at a time when bank transfers and credit cards were largely unknown and subscription to overseas publications was therefore problematic.

The Sky at Night changed all that. Despite the unsociably late hour at which it was transmitted and the clunky amateurishness of its sets and props, the eccentric enthusiasm of its presenter and the achingly beautiful theme music (Sibelius's "At the Castle Gate" from *Pelléas et Mélisande*, chosen by Patrick himself) ensured a loyal audience, and its monthly appearance in the schedules was eagerly awaited by those such as me. Key to its success was Moore's self-effacing lack of fussiness and a practical approach aimed at getting viewers to go out and do astronomy for themselves. In 2011 I was privileged to be a guest at the celebratory event at Patrick's home to mark its 700th episode, and at the time of his death in 2012, *The Sky at Night* held the record for the longest-running television series with the same presenter.

The program still runs in the UK with new presenters, but it has changed, and its original emphasis on practical observational astronomy has been much diminished. Those changes, along with the slicker approach to presentation, with high-quality graphics and incidental music (Patrick never allowed music during the program, as opposed to over the credits), are representative of a profound shift in how science as a whole is now presented and consumed in the UK media. Patrick Moore emerged as a personality-celebrity from the subject he presented. The latter came first. As he often said, "You look up . . . you can't help getting interested and it's there. We've tried to bring it to the people. . . . It's not me, it's the appeal of the subject."[2] Nowadays, there are many new popular science television programs and a new generation of skillful presenters. They are generally excellent, but the much greater emphasis now placed on the personality of the presenters, the huge budgets, and high production and entertainment values are a far cry from Patrick's conviction that astronomy could make its own case without a veneer of presentational polish and a cult of personality. We live in different times.

As was the case with many others, my first encounter with Patrick was as the result of a letter I wrote to him as a child and to which he responded quickly and encouragingly. How he found the time to reply to all who wrote to him, I will never know. Later I got to know him better via the Lunar Section of the British Astronomical Association (BAA), of which he was director from the midsixties to the early seventies. His invitation to me to join the committee of that section paved the way for a closer friendship that lasted until his death, an event that occurred during my presidency of the BAA. It fell to me to write his obituary for the *Journal of the British Astronomical Association*, a bitter task but one that allowed me to pay full tribute to his contribution to both the BAA and amateur astronomy as a whole (figure 8.1).[3] Patrick no doubt had his faults: he could be insensitive and bullish, and he projected a public persona that was not to everyone's taste. But to me, as to so many others, he was unfailingly supportive, generous, and kind—and quite unforgettable.

As a teenager in the early sixties, I joined the Liverpool Astronomical Society, a local society but one that had played a long and distinguished role in the history of British amateur astronomy, including acting as midwife to the

2. This quotation is from personal correspondence, but Patrick repeated the same idea on many other occasions. See, for example, *Patrick Moore: The Autobiography* (Stroud: Sutton, 2005), 28.

3. "Obituary: Sir Patrick Moore, CBE, FRS, FRAS (1923–2012)," *Journal of the British Astronomical Association* 123 (2013): 76–78.

FIGURE 8.1 Talking Moon: the author and Patrick Moore at the latter's home in Selsey, West Sussex, in 2009. Courtesy of Bill Leatherbarrow.

birth of the BAA in 1890. It was at the Liverpool Astronomical Society that I met Eric Strach, a local physician who went on to become a distinguished amateur observer of the Sun. Knowing of my lunar interests, Eric offered to introduce me to a friend who, in his words, "used to be an observer of the Moon." That friend was Harold Hill, one of the greatest lunar observers of the twentieth century, but who was by then more preoccupied with observing the Sun using his home-built spectrohelioscope (figure 8.2). Indeed, Harold had already passed on his main lunar telescope to Eric. That telescope had a distinguished provenance, having first belonged to the young and highly talented observer Samuel Morris Green, who was killed in action on D-Day at the age of 24, his first day of active service.

Harold did not suffer fools gladly, and he was distinctly out of step with the attitudes of the younger generation. I was young when I first met him, and probably foolish too, but he took me under his wing, setting me specific lunar observing programs, criticizing my efforts, and gradually coming out of "lunar retirement" to join in. From him I learned the value of methodical and cautious lunar and planetary observation, as well as the need to earn

FIGURE 8.2 *Left to right,* Richard Baum and Harold Hill having one of their rare face-to-face visits at Hill's listed eighteenth-century residence, Dean Brook House, near Wigan, in September 1993. Also present was an American lunar enthusiast, Bill Sheehan (not shown). Courtesy of Julian Baum.

one's spurs through a process of constantly acquiring experience, criticizing one's self, and developing skills. Unfortunately, I never acquired from him the supreme artistic talent that allowed him to produce such breathtaking (and painstaking) depictions of the Moon's complex topography.[4]

If Harold Hill brought the grit of criticism to my learning process, my third mentor taught me the importance of locating one's observational experiences in the historical context of observational astronomy. Richard Myer

---

4. See Harold Hill, *A Portfolio of Lunar Drawings* (Cambridge: Cambridge University Press, 1991).

Baum had an international reputation as both a skillful observer and a meticulous historian of solar system astronomy. Yet he had no formal education in that field, no academic affiliation, and was entirely self-taught. He was also possessed of a deep romantic regard for the mysteries of our neighboring worlds. Those mysteries, disclosed by generations of past observers, were slowly being dispersed by the discoveries made by spacecraft exploration, but Richard was always keen to understand, not simply disparage, the illusions of the past. We must not simply reap the harvest of modern scientific exploration without regard for the paths trodden by past telescopic explorers.

I count myself very fortunate to have had such guides as Patrick, Harold, and Richard as I took my early steps, but it is also important to consider the key role played by organizations such as the British Astronomical Association, the Royal Astronomical Society, and local societies in providing an essential social network for those engaged in what might otherwise become an isolating pursuit. For me, the most important of those by far has been the BAA, which I joined in 1965. By that stage I had read many books on astronomy, and the ones that made the most impression were the guides written by Patrick Moore. Works such as *Guide to the Moon, Guide to the Planets, Guide to Mars*, and so forth all contained an appendix on how to observe the subject and a description of what "useful work" could be done by the amateur observer. That phrase "useful work" certainly fired up my youthful imagination and provided the impetus needed to observe regularly in freezing temperatures; but the BAA, and particularly its observing sections, fleshed out for me the true meaning of that phrase.

From its formation in 1890, the BAA has had as its primary objective "the association of observers, especially the possessors of small telescopes, for mutual help, and their organization in the work of astronomical observation."[5] The primary means of achieving this has been "the arrangement of Members, for the work of observation, in Sections under experienced Directors." Take note of that word "work" again, for the BAA throughout its history has subscribed to the notion that the amateur astronomer is capable of making a significant contribution to knowledge. Of course, words such as "useful" and "significant" are relative. In 1890 there were comparatively few professional lunar and planetary astronomers, so those terms were largely

---

5. Quoted from the inside back cover of the annual BAA *Handbook*.

defined through the observational activities of amateurs. In lunar study, for example, amateurs were largely unchallenged by professional colleagues, and much of the work of mapping the Moon had been done by unpaid selenographers. Thomas Gwyn Empy Elger, the first director of the BAA Lunar Section, initiated an observing program that focused on the detailed mapping of selected areas of the lunar surface to establish the reality or otherwise of reported physical changes. This was continued by his successor, Walter Goodacre, who commented in 1933, "One of the chief sources of pleasure to the lunar observer is to discover and record at some time or other details not on any of the maps; it also follows that if in the future a map is produced which shows all the detail visible in our telescopes, then the task of selenography will be completed."[6]

But the devil lay in that very detail. Larger and larger aperture telescopes revealed smaller and smaller detail. Maps became bigger and bigger—and more and more cluttered and indecipherable. In that process, which overlooked the bigger picture, a true understanding of the Moon was lost. Moreover, amateur selenographers were generally not professional cartographers, with the result that their maps were often deficient and inaccurate in the sizes and positions of even quite large-scale features. They would certainly not do for the space age and the possibility of onsite exploration of the Moon.

When I joined the BAA Lunar Section in 1965, Patrick Moore had recently become its director. He certainly revived interest and inspired enthusiasm, and membership of the section grew significantly. Patrick, however, was a product of the old school of amateur selenography, having worked closely in the fifties with Hugh Percival Wilkins, perhaps the last representative of the age of great amateur Moon mappers. The trouble was that the emphases of that old school were now being eroded by the encroachment of professional lunar science and the results returned from spacecraft. What was the point in chasing down the finest lunar details revealed by a 6-inch or 8-inch reflector, when Ranger spacecraft were flying to the Moon, and orbiters were about to map it in detail? There was a vast chasm opening up between amateur and professional lunar study, and this led to some sharp disagreements between Patrick and some professional colleagues, especially in the United States. In the period between Wilkins's departure in 1956 and

6. W. Goodacre, "Fauth's New Moon Charts," *Journal of the British Astronomical Association* 43 (1933): 212.

Patrick's appointment in 1964, the Lunar Section had been led by a series of professional scientists who had tried to instill a more analytical approach. Ewen Whitaker had tried to introduce professional imagery with large telescopes as a basis for the section's work, and David (Dai) Arthur, who *was* a professional cartographer, had advocated an emphasis on the accurate cataloging of craters instead of the endless pursuit of finer detail. Both Whitaker and Arthur were subsequently recruited to Gerard Kuiper's lunar project in the United States and went on to work alongside some of the other contributors to this volume, playing important roles in mapping the Moon in the runup to the Apollo missions. Following on from Whitaker's directorship, Gilbert Fielder, a professional geologist, had advocated an observing program based on accurate charting of the distribution of different types of landform over the lunar surface as a whole.

It is probably true to say that Patrick remained unsympathetic to such approaches, and he took the Lunar Section that I had joined in a very different direction. While still insisting on the value of amateur telescopic mapping, particularly of the difficult areas near the lunar limb, Patrick's emphasis shifted to another old school preoccupation of selenography—the search for things happening on the Moon, or transient lunar phenomena (TLP).

Historical records of such phenomena go back several hundred years and the number of reported "changes" cataloged runs to several thousand. Clearly the incidence of such reports increased greatly following the invention of the telescope, and the notion of changes occurring on the Moon was particularly reinforced following Julius F. J. Schmidt's claim in October 1866 that the previously deep 5.6-mile crater Linné, once prominent on the Mare Serenitatis, had turned into a small whitish patch. Once thus established, the notion of a Moon where small-scale changes could still be observed was difficult to shift. It was further reinforced in the sixties when the traditional amateur territory of charting the Moon's permanent features was overtaken by spacecraft exploration, and amateur attention shifted instead toward the search for transient phenomena. The result was an explosion of interest and, it is fair to say, a sometimes uncritical acceptance of many instances of reported change—usually in the form of surface glows and discoloration, as well as flashes of light or apparent obscuration of details. Patrick Moore found some professional support for his TLP program in the United States, collaborating with Barbara Middlehurst, who had worked with Kuiper at Yerkes and then Tucson, and Winifred Sawtell Cameron, of the Goddard

Space Flight Center in Greenbelt, Maryland, to produce consolidated lists of reported events.

The ultimate apparent professional endorsement of amateur interest in TLP came with the establishment of NASA's Operation LION at the end of the sixties. The Lunar International Observers Network, timed to coincide with the early Apollo flights to the Moon, was a network of both professional observers and amateur volunteers whose task it was to monitor the lunar surface for "short-lived phenomena" during the astronauts' proximity to the Moon. Nothing conclusive was established, but Operation LION was an encouraging early example of pro and amateur collaboration in solar system astronomy.

At the time I was—and remain—rather skeptical about the reality of most TLP or the likelihood of surface changes of such magnitude as to be visible from Earth through amateur telescopes. So my own participation in LION was minimal, although I did my best to observe the Moon as regularly as UK weather permitted during the Apollo 11 mission. It is probably fair to say that from the late sixties through the nineties, something akin to a cult of TLP, which was of course really a cult of Patrick Moore, took possession of the BAA Lunar Section, and many reported events were almost certainly fanciful. That is not to dismiss all TLP out of hand. Clearly some are real, such as meteor impact flashes, which can and have been observed as confirmed events using amateur telescopes. These have accounted for only a small proportion of reported TLP, however, which for the most part have taken the form of strange glows, transient colors, and obscuration of surface detail. In many cases these might be explicable in terms of chromatic aberration, poor seeing conditions, or observer inexperience.

Whether such other forms of TLP occur on the Moon remains an open question and depends very much on how broadly one defines TLP. Fresh impact craters and other geological changes certainly happen, but they are likely to be small in scale and detectable only on spacecraft imagery rather than with Earth-based telescopes. The study of TLP remains part of the BAA Lunar Section observing program, but the emphasis today is on the reexamination of past reported events under repeat conditions of libration and solar illumination. This program, ably managed by Tony Cook on behalf of the BAA and the Association of Lunar and Planetary Observers in the United States, has brought a welcome sobriety to the evaluation of past reports of TLP.

By the midsixties I felt confident enough in my developing observational skills to take a full part in the work of the BAA Lunar Section. I developed a particular interest in craters such as Aristarchus that displayed prominent dark bands on their inner slopes, usually arranged in a radial pattern. The bands in Aristarchus had been observed by many classical observers from the mid-nineteenth century onward, and in 1955 Keith Abineri and Alan Lenham published a paper in the *Journal of the British Astronomical Association* that offered a categorization and distribution of 188 similarly banded craters, thus showing that Aristarchus was by no means the sole example. Lenham was yet another BAA observer—in fact, the first—who went to the United States to work with Kuiper. Some observers even thought that the bands might be subject to changes in extent and intensity. In the forties Robert Barker had argued that accelerating "evolutionary change" was taking place that spoke of the possible "growth of lowly vegetation on another world, 240,000 miles distant."[7] By the sixties the idea of vegetation on the Moon had largely receded, and my own studies showed no evidence of variability in the bands other than what you might expect under changing conditions of incident solar illumination (figure 8.3). But those studies did suggest that at least some of the bands appeared to occupy radial dikes in the crater wall. Aristarchus again was a good example of this. Images from spacecraft have now confirmed that radial bands are little more than evidence of scree that slid down slope after the impact process that formed the crater.

The first paper I read at a BAA meeting was on banded craters. The event was a meeting of the Lunar Section at Keele University on July 23, 1966 (the day on which the England soccer team, on their way to World Cup glory, beat Argentina in the quarter finals). I was eighteen years old, and to this day I don't know whether my nervousness on that day was down to anxiety at the thought of speaking before a large audience of elders or uncertainty about the outcome of the match.

Another Lunar Section observing project that seemed useful at the time was the charting of features at the lunar limb, which were difficult to observe because of extreme foreshortening, and some had not yet been properly imaged from the Orbiter series of spacecraft. One such project that I took part in was charting the large feature that Wilkins had named Caramuel (in honor of Juan Caramuel y Lobkowitz) but that the International Astronom-

7. Robert Barker, "The Bands of Aristarchus," *Popular Astronomy* 50 (April 1942): 195.

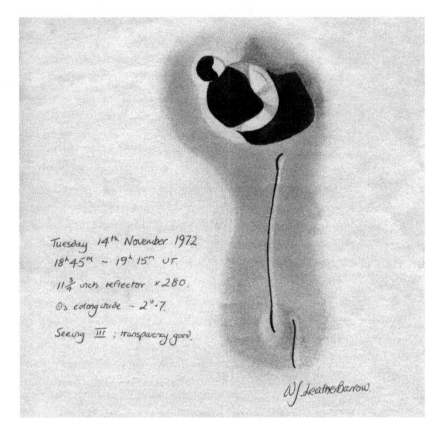

Tuesday 14th November 1972
18h45m ~ 19h 15m UT.
11¾ inch reflector ×280.
☽'s colongitude ~ 2°·7.
Seeing III ; transparency good.

WJ Leatherbarrow.

FIGURE 8.3 The banded crater Birt, drawn by the author. Courtesy of Bill Leatherbarrow.

ical Union had renamed Einstein in 1963; it lies on the extreme western limb of the Moon, and can be seen only during favorable librations. Einstein was particularly favorably presented on the night of November 8, 1965, when it was observed by many Lunar Section members, including me. A subsequent report appeared in the BAA *Journal*, but disagreements with professional colleagues in the United States over such fundamental issues as positional details and nomenclature exposed the gulf that was opening up between professional and amateur lunar study, exacerbating tensions between Patrick and American colleagues.

Less contentious but similar work was being carried out by Harold Hill, who was carefully charting those areas of the Moon's south polar regions that had been inadequately imaged by the Orbiter craft. Hill continued this work for decades to come. I joined in from time to time, but the skill levels required to disentangle and draw the elusive and intricate detail of such a foreshortened region exceeded those available to me, and it might be argued that the project eventually defeated Harold too, as Canute-like he tried to resist the encroachment of ever more comprehensive spacecraft mapping of our satellite.

In the autumn of 1966 I left my parental home in Liverpool and became a student at the University of Exeter, in the far southwest of Britain. On arriving there I made the acquaintance of J. Hedley Robinson, who at that time was director of the BAA Mercury and Venus Section. Hedley's benign and supportive influence did much to encourage my interest in planetary observation, and he quickly asked me to serve as secretary to his section. He took a particular interest in a long series of observations of Venus that I had made during the favorable eastern elongation of spring 1964, when the UK enjoyed an unusually sustained spell of clear weather. Hedley was of the view that my observations provided some support for recent ultraviolet imaging conducted over a period of sixty-eight days by Charles Boyer and Henri Camichel—the latter using the 60-cm reflector at the Pic du Midi Observatory—which suggested a rotation period of about four days for the Venus cloud layer. Hedley encouraged me to write up my observations in a paper for the BAA *Journal*.[8] My eyes have never been particularly sensitive to light at the blue end of the spectrum—and now, in my eighth decade, they are even less so. I have always found the markings on Venus to be extremely elusive and uncertain, so I am unwilling to stake a confirmatory claim to the findings of Boyer and Camichel, but Hedley's encouragement certainly spurred me on to further planetary work alongside my lunar interests (figure 8.4).

Apart from observation of Venus, both in integrated light and using color filters (the latter an approach enthusiastically embraced by Hedley), that work included observations of Mars to confirm seasonal changes in that planet's albedo markings and transit observations of visible features on Jupiter. Careful timing of such features as they transited the central meridian allowed one to calculate longitudes and eventually establish drift patterns—work that still goes on today, but in an entirely different way.

In 1970 I was elected for the first time to membership of the BAA Council, but regular travel between Exeter and London was not straightforward

---

8. W. J. Leatherbarrow, "The Rotation of Venus—Observations Made During the Eastern Elongation of Spring 1964," *Journal of the British Astronomical Association* 81 (1971): 177–80.

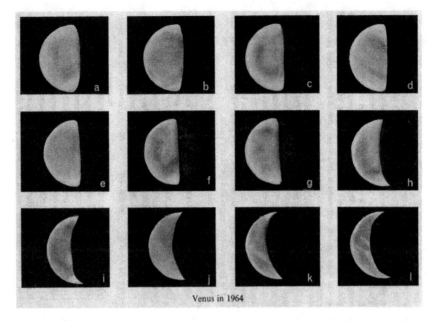

Venus in 1964

FIGURE 8.4 Some of the author's telescopic drawings of Venus in 1964, which seemed to support the evidence for a four-day rotation period in the atmosphere of Venus. Courtesy of Bill Leatherbarrow and the British Astronomical Association.

in those days. As a result I served just the one year, but that was enough to give me insight into how such organizations work and how they can give rise not only to collegiality, but also at times to intense factionalism. After my first meeting I was shepherded to one side by a group of senior members that included Patrick and Hedley and was inducted into what was described as the Polaris Club (named after a ballistic missile of the time). This turned out to be a sort of inner sanctum of council members selected by invitation only, who met in a nearby pub to "straighten out" the business of the formal council. It was great fun, and I returned to Exeter feeling rather elated that I had been deemed worthy of membership.

After the next monthly council meeting, however, I was again shepherded to one side, this time by a group of younger members, and was invited to join the Blue Streak Club (named after another ballistic missile). This was a rival gathering of young Turks keen to move British amateur astronomy and the BAA in a more progressive direction. Several of them went on to distinguished careers in astronomy, but at that time they numbered among those that Patrick was wont to dismiss as "serpents." The lines were drawn!

I decided to skip the third meeting.

Also in 1970, I moved to the University of Sheffield to take up what was advertised as a temporary lectureship in Russian studies. Forty years later I retired from the same university, having moved up the academic ladder, becoming in turn professor of Russian, head of the School of Modern Languages and dean of the Faculty of Arts. Shortly after my arrival in Sheffield, the then head of the Physics Department, Tom Kaiser (1924–98), decided to introduce astronomy as an undergraduate degree subject, something quite rare then in British universities. Tom was a distinguished space physicist, whose deeply held communist beliefs had given rise to a colorful and peregrinating academic career. He had previously worked in his native Australia and at Jodrell Bank under Sir Bernard Lovell, and in 1994 he was awarded the prestigious Gold Medal of the Royal Astronomical Society, the highest award the society gives, usually given in recognition of a lifetime's work. Tom brought several young professionals to Sheffield to staff the new degree courses, and he was keen to establish an observatory. Unfortunately, neither Tom nor his new appointments knew much about the practicalities of acquiring and using a telescope. Tom was the most approachable and democratic senior academic I had ever met, and knowing of my amateur interest in astronomy, he turned to me—then a very junior member of a different faculty—for advice on buying an instrument and a dome. Once the equipment was installed, he also invited me to speak to the first-year astronomy students on matters of practical observation. To this day I feel proud of the fact that, thanks to Tom, I can claim to have taught in both the arts and pure science faculties. Tom is long gone, but I could not help thinking of him when in 2016, the University of Sheffield awarded me an honorary doctorate of science (DSc) in recognition of my "shadow career" as an amateur astronomer.

In the late eighties and nineties, family commitments and the ever-growing pressures of my professional life took their toll, and I found myself increasingly withdrawn from active amateur astronomy, although my interest in the subject in general was undiminished. If the truth is told, I was also suffering from disillusionment, arising from the realization that the very successes of the space age that had so enthused me, coupled with the increasing professionalization of solar system science, were significantly eroding the value of traditional amateur activities. What was the point of painstakingly drawing and charting the Moon when orbiting spacecraft had mapped its surface so thoroughly? How could amateur telescopic drawings of, say, Mars compete with the detailed images returned from space? How "amateurish" earlier amateur efforts now seemed. Was there any longer a role for observers with their backyard telescopes?

That gloom was dispelled in a most dramatic fashion in the first decade of the new millennium, and most fortuitously for me, it coincided with my retirement from my academic post and the return of productive free time. Until this point the skilled and experienced eye at even a modest telescope could, under good seeing conditions, reveal more detail on the Moon and planets than could be photographed even with large professional telescopes. But the revolution that was ushered in with the advent of digital imaging and the wide availability of CCD cameras changed all that. Amateurs now had powerful new tools at their disposal. Solar system imaging in particular was transformed by the advent of high-speed video cameras, based on humble webcams, that allowed for "lucky imaging," whereby thousands of frames in a video file could be sorted by quality, aligned, and stacked to yield sharp and detailed images of our neighboring worlds. The resulting improvements were outstanding, even for new imagers such as me, who had limited processing skills (figure 8.5).

FIGURE 8.5 The "amateur's planet," Jupiter, imaged in two different eras. *Left*, in 1952 with the 200-inch Palomar reflector, showing a prominent Great Red Spot and a shadow transit of satellite Ganymede; *right*, in 2017 with the author's backyard 12-inch telescope. Courtesy of Mount Wilson and Palomar Observatories and Bill Leatherbarrow.

The digital-imaging revolution has, of course, completely revolutionized professional astronomy too, but for amateurs, the possibilities were transformational. Not only did it allow more detailed observation of the Moon and planets, but it greatly increased the overall efficiency of amateur work. One of Harold Hill's beautiful drawings of the difficult south polar regions cost him years of time and effort, with a single sketch easily being the work of an entire night at the telescope. Now the same complex areas can be recorded with much greater fidelity and higher resolution within a matter of minutes (figure 8.6).

The derivation of longitudes for Jupiter's spots and storms, which used to require painstaking effort, is now much simplified, since these can be measured directly from amateur images without the need for laborious timings of central meridian transits. This efficiency has helped to transform our understanding of how Jupiter's atmosphere behaves. Some amateurs have even been able to capture impact flashes as Jupiter's gravitational field sweeps in a passing asteroid. On Saturn, where storms and spots are less obvious, visual observers of the past saw them infrequently; nowadays they are detected regularly on amateur images. Amateurs have even imaged storms on the distant ice giants Uranus and Neptune.

For someone like me, whose astronomical interests were awakened by the romance of the space age, the planet Mars has always had a special fas-

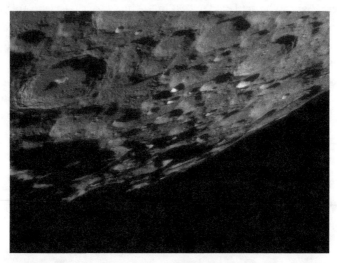

FIGURE 8.6 The Moon's south polar regions near Moretus, imaged by the author with a 12-inch telescope, May 13, 2019. Courtesy of Bill Leatherbarrow.

cination, and I am astounded at the fact that even an imager as limited as I am is now capable of capturing sufficient detail on that planet to permit systematic scrutiny of seasonal changes, albedo variations, and orographic clouds around the Martian volcanoes (figure 8.7). Who would have thought such things would be possible back in 1957?

The developments in imaging technology mean that amateurs are now able to provide high-quality ground-based support for professional space missions. Unhindered by the need to apply for time on professional telescopes, they can offer a continuous and comprehensive observational coverage not possible for their professional colleagues. Significant pro-amateur collaboration of this sort has been part of the Juno mission to Jupiter and of various missions to Venus. It will no doubt continue into the future.

The advent of the worldwide internet and the availability of powerful home computers have also encouraged the growth of pro-amateur cooperation. Numerous citizen science projects have attracted mass participation in the analysis of professional data. Moreover, imagery and datasets from spacecraft, available through online toolkits such as those on the lunar website QuickMap (https://quickmap.lroc.asu.edu/), now allow amateurs to manipulate sophisticated data to complement their telescopic observations.

For an amateur of my age, the speed at which technology evolves now can leave a sense of bewilderment and of being left behind, particularly when one sees what can be achieved with it in the hands of those who are younger and more adaptable. But then I remember that "there is no such thing as now," and that today's "now" will become tomorrow's "then." Provided that humankind survives the wounds it occasionally likes to inflict on itself, the technology available to the amateur astronomer will continue to evolve and improve. As an elderly child of the space age, I will not be around to see it, but that is the price you pay for having been a member of that unique and privileged generation that witnessed firsthand the initial voyages of discovery to our neighboring worlds.

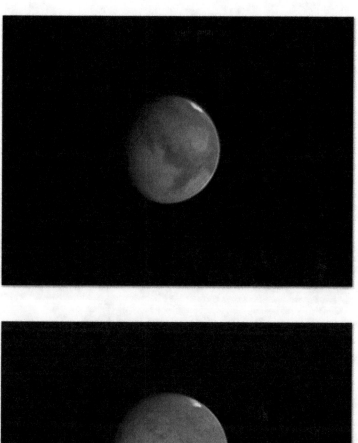

FIGURE 8.7 Digital images of Mars by the author with a 12-inch telescope. *Top*, August 31, 2020; *bottom*, September 15, 2020. Courtesy of Bill Leatherbarrow.

Growing Up with Apollo

YVONNE JEAN PENDLETON

Everyone has touchstone moments in their life from which they see their path take a twist or turn. Looking back, I believe the Apollo launches in the late sixties provided the most significant touchstone moments for me, with a clear delineation between the time before I fell in love with space and every day thereafter. I was born in Key West, Florida, in 1957, just ten days before Sputnik orbited the Earth and changed the world. Sputnik spurred on the U.S. space program, and by the time I was old enough to care, Apollo missions were launching frequently. Some of the Apollo launches could be seen, on a clear day, from my backyard. My father died when I was thirteen, but sometime before that, I have a vivid memory of standing in our backyard, holding his hand, and looking up as one of those rockets soared out of sight. I told my dad that when I grew up, I would "work for NASA and study the stars." It seems incredible to me that I would have said such a thing at that age, but others remember it as well as I do. Perhaps more incredible is the fact that those words became my reality.

As I began writing this piece in 2021, I passed the forty-two-year mark of civil service work with NASA, the government agency that has meant more to me than I can adequately express to a depth that I am, frankly, slightly embarrassed to admit. The circuitous path that took me to the right door at the right time—repeatedly—and the dogged determination that was born on that day as a little girl looked up through cloudless skies resulted in a career choice that has brought me deep joy and satisfaction. I hope I have contributed in meaningful ways to science and to others along this journey, but if so, those are just the cherries on top of this delicious ice cream sundae.

I study stars at the earliest moments of their creation, as well as the ice and dust that compose the environments from which they form. The planetary systems that emanate from the smallest particles and planetesimals that come together around those stars can provide safe harbor to materials necessary to form life, at least on this planet, and that fascinates me. I have followed the origin and evolution of interstellar dust and ices, with a special interest in the increasing complexity of carbon-based molecules. As an astrophysicist and planetary scientist at NASA's Ames Research Center in the heart of Silicon Valley, California, I was given the opportunity to work from some of the world's finest telescopes to gather the data necessary to probe stardust. It has been an exciting and rewarding journey, and while I officially retired from NASA in October 2021, I affiliated with a university in 2022 and have no plans to ever retire from scientific research. "Choose a job you love, and you will never have to work a day in your life," reputedly said by Confucius, sums it up nicely.

Formative Years

My father studied electrical engineering in college and had his own TV repair shop for many years. I remember our garage was always filled with rows of TV picture tubes. He also taught engineering technology courses at the naval Fleet Sonar School in Key West, and he loved to play chess. In his spare time, you could always find him on the other side of the chess board, which was seldom put away. He would play the "Master's games," studying their moves carefully. He taught me to play chess over several years, starting when I was about five. His strategy was wonderful—he began by having me play him with all his pawns and just his king. When I could successfully beat him at that level, he would add one more piece to his side. In all the games we played, I only beat him twice when he had all his pieces. The glowing pride I felt when I said "checkmate" didn't hold a candle to the pride I saw in his eyes as he said, "Good game, Sweetheart." My mother was from Naples, Italy. They met during World War II, fell in love, and later she bravely came to the United States on a ship as a "war bride" at the young age of twenty-one. She had loved math in school, but the war made it impossible for her to continue after high school. She became a successful bookkeeper at banks in Washington, D.C., Miami, and Key West, and in the early days, she did so before she could even speak English.

My own pathway resulted from a series of good choices and a serious amount of good luck. Having taken as many science classes as possible during high school, including third-year chemistry, and participated in extracurricular activities, such as applying for the Skylab student experiment contest when I was in ninth grade (mine was not selected), I was "grooming" myself for NASA, even though I had no idea how to actually get there. I went to an excellent high school, Druid Hills High School, near Emory University (Atlanta, Georgia), so that was a helpful start. The ability to attend Druid Hills High was the result of one of the toughest things that happened to me as a child, though. My father died right before Easter the year I was in eighth grade. I remember thinking, that very night, that forever onward I would benchmark whatever happened to me against this difficult loss, and such has been the case. I had lived in Key West until my father's death, and it was all I had ever known, but when my mom gave me the choice of staying or moving elsewhere, I didn't hesitate to choose Atlanta, Georgia.

My sister, Olga, nearly ten years older than I, was in graduate school at Emory University, where she earned her PhD in biometrics and statistics. We were very close, and I credit her for the good decisions I made (most of the time) while growing up. Olga was the reason I began working harder in grade school, and once I did, I remained a very good student. In some ways, she was a third parent, and the one who absolutely refused to let me take the easy road (academically). When I came home with a report card peppered with Cs in the fourth grade, it was my sister who made an appointment with my teacher and went, on her own, to tell the teacher not to accept this from me. She knew from experience that I could do better, because, as she tells the story, when I was about five years old, I was already doing long division. She discovered this one night when she was babysitting me and was tired of my bugging her while she was doing her math homework. In frustration, she opened one of her desk drawers and found an old seventh-grade math workbook with long division problems—many pages of them. She took some time to explain how to do a few problems and then "ordered" me to try to do some on my own. Well, I adored my sister, and I did everything she told me to do (most of the time). So, sometime later, after she had completely forgotten I was there, she turned around to find me still at work on the math problems. She was so proud that I had the attention span to sit there and at least "try" to do them, but she was absolutely shocked when she checked them and found I had done them correctly. Now, I figure I must have made some mistakes, but she claims I did not, and I still believe her (most of the time). Anyhow, this is what she told Miss Ezell, my fourth-grade teacher, and after that I was definitely expected to shape up. It didn't happen overnight. I had been the queen of "chatter box lane" in third grade, so it was going to take some work to steer this boat in a new direction. By sixth grade, however, my report cards show all As and Bs, then later all As, and I actually did not make another C until my sophomore year of college (in a political science class, of all things). So, it will come as no surprise to anyone that my PhD thesis dedication reads, "To my sister, Olga Pendleton, who taught me to reach the unreachable stars."

At Druid Hills High School, I was given many opportunities, but the one I cherished most was the ability to spend quality time at the nearby Fernbank Science Center during my senior year. I was taking third-year chemistry but was the only student in the class. When my teacher, Mrs. Strong, wondered what I wanted to do as an independent study course, I asked if I could work with a chemist I knew at the Fernbank Science Center, Ralph Buice. She talked to him, and they agreed, so from then on, I spent the last period of the day at Fernbank, studying proteins in egg albumin. I would often spend my evenings among the stars (in the Fernbank planetarium or observatory). Over the previous two years, I had taken a couple of night courses there, one in mass spectroscopy and another in astrophotography, and I was a frequent visitor to their planetarium and public viewing nights. The astronomers on staff were very encouraging in my pursuit of "all things space related." In 1973, I had a memorable fight with my mom because I wanted to go with a Fernbank-sponsored group to see the comet Kohoutek, and she refused. Her concerns were that we would have to drive quite a distance (to get out of the Atlanta lights), I was the youngest and the only girl in the group, and she didn't know "those men" (the astronomers, Rick Williamon and Ralph Buice). As a mom myself, I might have had the same concerns, although I would have said, "yes, but only if I can come along." When the comet turned out to be somewhat of a dud, I was only slightly less angry.

My mother and I had another fight that was astronomy related when I was fourteen. I really, really wanted the book *Astronomy* by Donald H. Menzel.[1]

---

1. Menzel, *Astronomy* (2nd ed., New York: Random House, 1970).

It cost thirty dollars, and I was diligently saving most of my allowance each week to buy a telescope. I begged her to buy that book for me, but thirty dollars for a book I would, in her words, "never read" was too much to spend. In the end I wore her down, but to my credit, I *did* read the book, and today it sits proudly on my bookshelf and reminds me of my tenacity. My mom ended up becoming one of my best friends and a proud supporter when I was older, but we sure didn't see things the same way when I was a teenager.

Books are so important at every stage of one's life. I developed my deep love of reading at a very early age, but not because I was especially curious. This was when I still lived in Key West, Florida, and our house, like all the others in "old town," was not air conditioned. The library at the end of my block, however, was. As a consequence, I spent most of my summers there, and I always won the "who can read the most books" contest. I felt a little guilty getting whatever the prize was because I wasn't even trying. When you spend that much time in a lovely library, surrounded by really good books, this is just bound to happen. I am sure my considerable difficulties passing the softball throw in the Presidential Physical Fitness tests in PE class were directly related to the way I chose to spend my summer days, but reading all those books year after year developed a deep love of the written word and increased my reading ability by several grade levels. I still can't throw a softball very far, but I can still get lost in a good book.

After moving to Georgia, sometime during my ninth-grade year on a church-sponsored trip, one of our chaperones sat next to me and found out I really liked "all things space related." Turns out he was a junior at Georgia Tech (where I would later end up) and was majoring in nuclear physics. Sometime after that trip, he brought me his copy of William A. Fowler's *Nuclear Astrophysics*, thinking I might like to have it.[2] It was too hard for me to understand at that age, but I kept that book. I still like the fact that such a cute older boy gave it to me.

Speaking of boys, I have a confession to make. The telescope I was dutifully saving for in my freshman year did find its way to me, but I didn't have to buy it. By the end of ninth grade, I had a steady boyfriend who "loaned" me his. He was the first guy who shared my interest in the stars, and we dated all through high school; hence, the telescope was essentially mine. I spent the

---

2. Fowler, *Nuclear Astrophysics* (Philadelphia, Pa.: Memoirs of the American Philosophical Society, 1967).

telescope money I had saved on clothes, as many sixteen-year-old girls in the South, at least in the early seventies, would have agreed was the wiser choice. I was definitely a nerd, but not so much so that I didn't care what the other kids thought. I remember covering my textbooks with brown paper in my junior and senior years of high school, so that the "popular" kids wouldn't know I was taking advanced science classes. I am fairly certain I was one of the few girls with subscriptions to both *Sky & Telescope* and *Seventeen* magazines.

A significant opportunity came about during the summer of my tenth-grade year. My first chemistry teacher, Mrs. Reed, suggested I apply for a National Science Foundation Summer Chemistry Program at Georgia State University. I was selected and I thoroughly enjoyed the ten-week experience. The love of astrochemistry I have today harkens back to the interest that was developed during that program, and I am so grateful to Mrs. Reed for seeing the potential in me that she did.

When it came time to select a college, I stayed fairly close to home because I didn't want to leave my mom alone for long periods. My older sister had earned her PhD by then and moved away, and I was keenly aware that I was the center of my mom's world (she never did remarry, but she later moved to where her sisters lived). I didn't want to immediately go to Georgia Tech, because that would mean living at home, rather than in a dorm, since it was so nearby. I really yearned to "go away" to college. So, I went to a beautiful small private school, Berry College, in the north Georgia mountains about sixty-five miles from where my mom and I lived. Although I spent only my freshman year there, it was transformative in many ways. I am still in awe of the beauty of that campus, more so than any other I have seen. I admire the strong woman who started the Berry schools and her reasons for doing so. I have several books about Martha Berry on my bookshelf today. Being a private school, Berry was expensive (which is ironic, since Martha Berry began it as a way to educate anyone who was willing to work hard), but I had a good scholarship, and they had a strong work-study program. My job could not have been more perfect. Because of my chemistry background, I was assigned to prep the chemistry labs for the introductory chemistry classes. After the first quarter, I was also invited to tutor the students struggling with the course. One of my chemistry teachers, Barbara Abel, and one of my math teachers, Robert Catanzano, especially encouraged me. At the end of my freshman year, though, I realized I needed to transfer to Georgia Tech if I ever wanted NASA to "find me." I do not know why I thought NASA would come looking for me, but after my freshman year at Berry College, I transferred to Georgia Tech, in the fall of 1976. What happened next was shocking.

One of the first things I did when I got to Georgia Tech was visit the job placement office and ask which NASA centers had recently come to recruit there. I was so disillusioned when the woman behind the counter looked at her records and sadly said, "Honey, NASA hasn't been here in *years*." She really emphasized years. Feeling somewhat rudderless, I left and wandered around looking at the buildings. How could this be? At the time, I was a chemistry major, but when I saw the Guggenheim building, which housed the School of Aerospace Engineering, bells started ringing. I changed my major that day, not because I liked airplanes, but because my new major had the word "space" in it. I figured if NASA were to come back, they would want aerospace engineers. I plugged away at the never-ending calculus requirements, the fluid mechanics and structures series, the electrical engineering and mechanical engineering courses, and two years later, in the fall of my senior year, every NASA center I had any interest in seeing came to recruit at Georgia Tech!

I was simply in the right place at the right time to capitalize on the recruitment visits NASA made to Georgia Tech in the fall of 1978. And they definitely wanted to interview aerospace engineers, so once again, good luck and good choices carried me forward. The entire recruitment experience was rather heady, and soon I had offers to work for NASA centers I had long admired (especially the Johnson Space Center in Houston, which is where I thought I most wanted to go), a few others, and one of NASA's smallest research centers, Ames, about which I had never heard. I was able to visit these as well as many aerospace engineering companies across the United States that fall, and in the end, I selected the offer that paid the very least, because it offered the very best next step in my education. NASA Ames recruited me with the offer not only to start me out as a GS-7 civil servant (at a whopping $16,900 per year), but also to pay my full tuition to Stanford University, if I could get in. They made it clear that they could not help me do that in any way—I had to get an offer from Stanford solely on my merits. This was to pursue a master's degree in aeronautics and astronautics, and I accepted the challenge. After applying to Stanford, I continued interviewing with the other companies, and I had the NASA Johnson Space Center offer as my backup plan, but by then I had already been bitten by the bug that infused me with the desire to learn more. I knew that a Stanford education would take me places I had yet to envision, but what I did not see coming was the complete career change working at Ames would cause me to take in the near future.

I clearly remember getting my acceptance letter from Stanford and making the call to the Human Resources Department at Ames within one minute of ingesting the happy news. I graduated from Georgia Tech in June 1979 and began working at NASA Ames in July. My "job" was to work part time and go to Stanford part time. My NASA work was in the Space Science Division, where I was exposed to myriad interesting research activities and solar system exploration missions. I remember seeing Carl Sagan, as he would visit Ames colleagues, and a few years later, he would actually come to visit me, too. I have letters Carl wrote to me about my research, and they are among my most prized keepsakes from that era. He made it a point to encourage young scientists, and I made it a point to follow that example when I got older.

The NASA Ames Research Years

By the time I finished my master's degree at Stanford in aeronautics and astronautics (1981), my interests had veered completely from the aerospace engineering side to basic research in astronomy and astrophysics. I can thank my first mentor at NASA Ames, David Black, for this course change. David and I worked on the three-body problem, concerning the stability of a planet orbiting a binary star system. David is a first-class theoretical physicist, and the years I worked with him (while also pursuing my master's degree) provided a dramatic contrast to the engineering courses I was taking. David allowed me to grow and discover my own interests by introducing me to observational astronomers Russ Walker and Larry Mertz, who were building a speckle interferometer at the nearby Lockheed plant. My first observing run was with Larry and Russ, using the NASA Mt. Lemmon 60-inch telescope and their speckle interferometer to image binary star systems in 1980. It was there that I fell in love with observational astronomy, although David and I continued to work together on our theoretical study a few more years. I am very proud of the paper we published together in the *Astronomical Journal*, and I am exceedingly grateful for the observing experience that pointed the way to my future career.[3]

---

3. Y. J. Pendleton and D. C. Black, "Further Studies on Criteria for the Onset of Dynamical Instability in General Three-Body Systems," *Astronomical Journal* 88 (1983): 1415–19.

Knowing I was leaning toward observational astronomy, David introduced me to Michael Werner, who had recently come to Ames as the project scientist for what was then called the Space Infrared Telescope Facility (SIRTF). After launch, SIRTF was renamed the Spitzer Space Telescope. Mike is an incredible infrared astronomer, and he undoubtedly taught me most of what I know about the field. After finding out more about the observational programs Mike had under way, I asked to work with him on observations of ice in star-forming regions to see if infrared observational astronomy was right for me. My first observing trip to Mauna Kea, Hawai'i (circa 1983), revealed that I had found the right fit. Infrared observational astronomy was what I most wanted to do. I had also developed the utmost admiration and respect for Mike and knew this was the person with whom I wished to pursue my PhD. To this day, I am so grateful for the opportunity to have worked with him. I applied for and was awarded a NASA graduate student fellowship, which fully paid for my PhD studies in astrophysics (completed in 1987) at the University of California, Santa Cruz (UCSC). I did this while continuing to work at Ames, and my work later became the basis for my PhD thesis. My adviser at UCSC was David Rank, who knew more about infrared instrumentation than anyone I have since met. He was wise, calm, and flexible, the latter being a quality of utmost importance to a graduate student trying to get two institutions to work together. David Rank could see the benefit of interlacing my work at Ames with his mentorship at UCSC, so I continued to obtain most of my data from the NASA Infrared Telescope Facility on Mauna Kea. Rank and I also observed from Lick Observatory and from the Kuiper Airborne Observatory a few times. After finishing my coursework, I passed my preliminary exams and was advanced to candidacy. My PhD thesis developed into a joint observation and theoretical topic on the properties of dust grains in two star-forming regions. In addition to David Rank, I consider my advisers to have been Mike Werner and Xander Tielens (both at NASA Ames). A. G. G. M. (Xander) Tielens was a postdoctoral fellow at Ames during this time, and it was my good fortune that he decided to take that position. Coming from Leiden University, he brought a unique capability to form theoretical, observational, and laboratory chemistry insights. The complexities of this universe are made clearer because of his work and his outstanding ability to teach others, and he is still one of my closest colleagues.

My observational work to this point had included near-infrared polarimetry studies of infrared reflection nebulae and spectrophotometry of the water band ice absorption seen in those regions.[4] After graduating from UCSC, I interacted much more with the astrochemistry group at Ames, becoming interested in the infrared spectroscopy of irradiated ice mixtures, which produce spectra similar to hydrocarbon absorptions that had been detected along sightlines toward the galactic center. Scott Sandford, Kris Sellgren, Lou Allamandola, Xander Tielens, and I worked to match the synthetic spectra to the observed data we and others had collected.[5] This soon became my primary area of focus, and I led observing runs and analyzed spectroscopic data collected along multiple lines of sight through our galaxy, to compare the strength and profile of the absorption bands. We found that similar chainlike lengths of CH_2 and CH_3 hydrocarbon groups are present in the intervening dust throughout the diffuse interstellar medium, and the amount of dust along each sightline corresponds to the strength of the bands present. Given the ubiquity of these hydrocarbons and the life cycle of dust, our observations strongly suggested that these fundamental building blocks become incorporated into the clouds from which stars and planetary systems form.

Over the course of my career, I have studied the origin and evolution of interstellar dust, ice, and organic molecules, beginning with their initial creation and following the changes that occur as they are exposed to various space environments. The processing of simple atomic and molecular materials containing carbon, hydrogen, oxygen, and nitrogen creates complex organic molecules. I have focused primarily on regions outside our solar system but within our galaxy, although another galaxy did alter my thinking in an exciting way (more on that later). In recent years, I have become fascinated by the chemistry of the early solar system and want to understand how it compares to that of other forming exoplanetary systems. Looking back over forty years, I can see that this is a logical continuation of my earlier studies, because the composition of interstellar dust in the diffuse interstellar medium (the region of space between star-forming clouds), plus the ices and organic molecules that form in dense molecular star-forming clouds,

---

4. Y. J. Pendleton, M. W. Werner, R. Capps, and D. Lester, "Infrared Reflection Nebulae in Orion Molecular Cloud 2," *Astrophysical Journal* 311 (1986): 360–70; Y. J. Pendleton, L. J. Allamandola, A. G. G. M. Tielens, and M. W. Werner, "Studies of Dust Grain Properties in Infrared Reflection Nebulae," *Astrophysical Journal* 349 (1990): 107–19.

5. Y. J. Pendleton, S. A. Sandford, L. J. Allamandola, A. G. G. M. Tielens, and K. Sellgren, "Near-Infrared Absorption Spectroscopy of Interstellar Hydrocarbon Grains," *Astrophysical Journal* 437 (1994): 683–96.

provide the inheritance available to new star and planetary systems. In our own solar system, there is evidence of primitive material that has remained relatively unaltered over the past 4.5 billion years, although much of it has been processed by energetic sources and the dynamical environment. I look for chemical signatures in these different regions and follow the chemical changes to deduce what has happened. My work is like that jigsaw puzzle one has in the living room of a comfortable house. You ponder and try different angles until, one day, the piece you have long been staring at really fits.

My first "ah ha!" moment came while I was attending a seminar given by Sherwood Chang at NASA Ames, sometime around 1991. It was a casual lunchtime seminar, and he was using viewgraphs—acetate films you would put on a lighted projector. He was discussing the organic material in the Murchison meteorite, which is an important carbonaceous chondrite that fell on September 28, 1969, in Australia near Murchison, Victoria. Because it was an observed fall, remnants were recovered quickly, and many have been well preserved. As Sherwood presented his talk, I saw that the organic signatures in the infrared spectra of this meteorite looked strikingly similar to the organic signatures in my observations of interstellar dust. The excitement of what a comparison between these might reveal grabbed hold of me and has never let go. Could the material responsible for the organics in the tiny dust grains of the diffuse interstellar medium be responsible for the organic signatures in the Murchison meteorite? After the talk ended, I asked Sherwood for those viewgraphs and quickly made my way first to my office and then to the Xerox machine. With some creative xerography (a course someone really should have taught back then, as it was how a lot of discoveries were made), I was able to stretch the scales until they were comparable. Then, holding the two figures up to the light, I could see what I had already suspected. The hydrocarbon absorption bands seen in the meteorite and in the interstellar dust exactly coincided with one another. While this was not clear evidence that the interstellar dust survived incorporation into the parent body of the meteorite when that body formed long ago, it was a possibility (and still is). I would spend the rest of my career trying to understand the composition of those organic bands and their histories.

Soon after that seminar, I had my first conversation with Dale P. Cruikshank, a man eighteen years my senior and someone already so well established in the field that I was quite intimidated to speak to him. I did not want to bother him with what might be my naïve wonderings, but I needed to know more about the solar system, and he was our resident expert. Dale had come to NASA Ames in 1988, but until now, we had only been introduced and spoke very briefly then. I finally went to his office and knocked on that door. He not only told me all about meteorites and began my quest to study the solar system, but also gave me good career advice, as I was just starting down a path he had already ventured. The advice he gave me, which I followed to the letter, launched my career. I have shared this same advice with other early career astronomers, but I don't think any of them took it quite as seriously as I did. Perhaps it had to be given by a giant in the field to be convincing, but it sure worked well for me. In case it still holds value for anyone reading this in the future, I will tell you what he said.

First, he said that in addition to publishing at least two scientific papers per year in peer-reviewed journals, I should find a way to present my work at a minimum of two professional meetings each year, whether NASA paid for my travel or not. Second, and perhaps most significant, because I never would have done this otherwise, he encouraged me to write a popular science article summarizing my work thus far and to get it published in *Sky & Telescope* or *Astronomy* magazine. I had no idea how many professional astronomers read *Sky & Telescope* each month, but it turns out to be a significant number. Third, I should volunteer to organize a meeting in my subject area, so that I would get to know the researchers in my field, and they would get to know me. Fourth, and this point he really stressed, I should commit to publishing the proceedings from that meeting within one year after it took place. "No dilly-dallying, like so many people do after these conferences occur!" Over the next few years, I did each of those things, many with his help and encouragement. The results included our jointly authored paper "Life from the Stars?," which made the *cover* of *Sky & Telescope* (figure 9.1), the "From Stardust to

FIGURE 9.1 Yvonne holding a piece of the Allende meteorite and the March 1994 issue of *Sky & Telescope* magazine, where the article "Life from the Stars?," by Pendleton and Cruikshank, made the cover, with the headline "Molecules in Space: The Key to Life?" Courtesy of Dale P. Cruikshank.

Planetesimals" Astronomical Society of the Pacific conference I organized (June 1996), and the ASP Conference Proceedings #122 from that meeting (published in 1997), plus the scientific papers and conference talks (including my first invited talk in 1992) that allowed me to develop meaningful relationships with colleagues all over the world.

Somewhere between all this scientific activity, somewhere in the Kuiper Belt, we fell deeply in love. It has been the most romantic journey I could have hoped for, as we worked side by side all these years in different, but related, fields. Dale is giving, kind, and generous with both his time and wisdom. He not only raised three boys of his own, but also helped raise my son and daughter from a previous marriage. We married in 1996, and these past twenty-five years together have been wonderful. Our mutual interest in the origin and evolution of organic molecules in space carried us like comets through the cosmos. Today, as I look back on the work we have done separately and together and think about the exciting results to come from future missions, I feel energized. Dale and I have now co-authored several scientific articles and worked together at telescopes on Mauna Kea (Keck Observatory and the NASA Infrared Telescope Facility; figure 9.2). We look forward to getting data from the James Webb Space Telescope (JWST) that will further tie the chemistry of the interstellar medium to the compositions of planetary bodies in the solar system.

FIGURE 9.2 Yvonne during an observing run to study Kuiper Belt objects, Keck Observatory, Mauna Kea, Hawai'i, 1996. Courtesy of Dale P. Cruikshank.

The second "ah ha!" moment of my career came about thanks to another close colleague for whom I have great respect, Tom Geballe. Tom worked on the telescopes atop Mauna Kea, Hawai'i, and I met him sometime in the early eighties. We have had the good fortune to collaborate on proposals and observations over the years, and just finished a new paper together this year.[6] Tom knew of my work on the near-infrared hydrocarbon absorption bands in the diffuse interstellar medium of our galaxy, and one night at the United Kingdom Infra-Red Telescope, while he was searching for the H_3^+ molecule in another galaxy, he recognized what might be "my" features. He contacted me and either sent me the data or faxed over the spectrum—I can't remember which happened first. I remember that I calculated the wavelength shift due to the galaxy's distance while waiting for the fax to arrive. When I overplotted the spectrum onto the data from our galaxy, correcting for redshift, the subfeatures of the aliphatic (chainlike) hydrocarbon absorption bands fit beautifully. So now we had the same structures appearing not only along several different sightlines through our own galaxy, and in the carbonaceous meteorite, but also in a nearby galaxy.

Like every astronomer I know, I have had the benefit of learning from various colleagues in many different fields. Each of us really does "stand on the shoulders of giants." Lou Allamandola is another "giant" in my journey-of-learning book. Lou is an extraordinary chemist who came from Leiden University to NASA Ames and created an astrochemistry lab that became a world-class facility. He is as patient and kind as he is brilliant, and he taught me a lot about organic chemistry. Together, we spent more than two years gathering, analyzing, and comparing observational data to laboratory results from experimentalists all over the world (about a dozen different lab groups) to understand the hydrocarbon chemistry we were seeing in the interstellar medium. We co-authored a paper published in the *Astrophysical Journal* that has, surprisingly to me, not yet gone out of style.[7] I am honored that Lou considers this "one of his top four best papers" (he has written hundreds of papers), but I am even more honored that he remains one of my dearest friends.

6. T. R. Geballe, Y. J. Pendleton, J. Chiar, and A. G. G. M. Tielens, "The Interstellar Medium Toward the Galactic Center Source 2MASS J17470898–2829561," *Astrophysical Journal* 912 (2021): 47–59.

7. Y. J. Pendleton and L. J. Allamandola, "The Organic Refractory Material in the Diffuse Interstellar Medium: Mid-Infrared Spectroscopic Constraints," *Astrophysical Journal Supplement Series* 138 (2002): 75–98.

In recent years, I have focused on the increasing complexity of the organic molecules that are observed in dense molecular clouds, some prior to the onset of star formation. The goal of studying the simple basic materials is to understand how greater complexity occurs, which leads to my most recent "ah ha!" moment. On January 1, 2019, the NASA New Horizons mission revealed the infrared spectrum from the surface of Arrokoth, a tiny body beyond Pluto in the Kuiper Belt. The spectrum clearly shows methanol ice, and on seeing this, I was overcome with excitement, knowing that methanol ice (CH_3OH) is a key component in the development of more complex organics in dense molecular clouds. We have long wondered to what degree newly forming protoplanetary disks might inherit such a key ingredient. With Dale and other colleagues, I participated in the first discovery of methanol ice on a solar system body (asteroid 5145 Pholus), which we published in 1998. This made the discovery of methanol on tiny primitive Arrokoth, which is farther from the Sun than Pholus, all the more exciting. Discovering methanol ice on Arrokoth has added significance because, unlike most solar system bodies, where larger planets have caused their orbits to change with time, the dynamical orbit in which Arrokoth (and other cold classical Kuiper Belt objects, or KBOs) currently resides is most likely the orbit in which it first formed. Furthermore, the region of the outer solar system where the cold classical KBOs orbit is relatively benign, implying Arrokoth is unlikely to have been heavily bombarded by many larger bodies over the lifetime of the solar system. Therefore, the methanol we see on Arrokoth today might be a remnant of its original composition, and if so, we have a unique insight into the composition of our own solar system, at a distance of forty-five astronomical units from the Sun, 4.5 billion years ago. The JWST will not be able to study tiny Arrokoth, but it will be able to see larger KBOs in similar orbits. So, while we will not learn more about Arrokoth itself, we will learn about the family into which it was born. Without the discovery of methanol ice on Arrokoth, we might not have placed such a high priority on the study of other KBOs in that particular region, and such studies will no doubt lead to insights into how planetary systems chemically evolve. Alan Stern, principal investigator for the New Horizons mission, deserves more credit than he will ever be given, for envisioning and seeing this mission through and for selecting the outstanding team members that have made it successful. What we have learned about Pluto would have been astonishing enough, but adding to it the insight we now have into the Kuiper Belt makes this a truly amazing mission.

The NASA Leadership Years

I have held several leadership positions within NASA, although I never applied for a single one. I loved research and didn't want to go into management, despite being asked to consider it over the years. In early 2006, however, I answered the call to go into midlevel management and became the first (and so far, the only) female division chief of the Space Science and Astrobiology Division at Ames. This is the same division where I began as a graduate student in 1979. I made this leap suddenly, when asked, because the very excellent division chief at that time, Michael Bicay, had been swiftly moved up the ladder to the directorate level, and we were all in fear of having some outsider step in who did not appreciate the culture of our wonderful organization. I naïvely thought this "acting" position would be just a temporary appointment—on that I was right, but it didn't mean I would return to my scientific research anytime soon.

I did not expect to like being a manager, and the timing of this step was especially worrisome because I had just won time on the Spitzer Space Telescope with a proposal for which I was the principal investigator. But almost immediately, I realized I really enjoyed the opportunity to make life better for the scientists in our division, and I seemed to have a knack for it. We had all suffered under the transition NASA was making to full-cost accounting, so I promised them we would put the *fun* back in *dysfunctional*. Very soon the acting position was made permanent, and I was catapulted into the high ranks of the senior executive service, but by then I was ready to embrace this new challenge fully. The longtime deputy division chief, Mark Fonda, continued to oversee the financial down-and-in aspects, and I began the frequent travel to NASA Headquarters to address our funding issues and to advocate for the great work our scientists were doing.

Our scientists were drowning under the weight of writing so many proposals to bring in their own funding, which full-cost accounting required that they do. This was a self-made internal NASA problem that needed (and still needs) to be fixed at a much higher level, but in the meantime, I looked for creative stopgap measures. We did not have enough money to hire additional administrative support, so all the areas of developing proposals, such as budget sheets, figure preparations, references, and so on, had to be done by the scientists in addition to the proposal content. Nowadays, this is common practice, and we all do it because we have to and because the tools to enable it got a lot better, but back then, these proposals were still being put together piecemeal, and those were time consuming tasks you really could have someone else do if you had sufficient clerical help. This was an obvious area where student support would have been an ideal solution, but well-meaning rules were in place that prohibited hiring science students to do administrative work, even if it would be useful for them to find out how funding proposals were created. And though we didn't have enough money to hire a full-time administrative assistant, we did have enough to hire a few part-time students. My solution was to hire business majors, rather than science majors, and this simple fix actually worked. We found a handful of really talented and hardworking local business students who wanted the experience, and we simultaneously reduced the stress level of the scientists, who could now focus on the content of their proposals. I say "we" because upon taking the job as division chief, I immediately recruited one of my dearest friends and chemist colleagues, Max Bernstein, to be the associate division chief. We figured that together, we could do the job and still maintain our science work at some level, and that is exactly what we did. The win rate on proposals actually increased that year, and from what we could tell, the scientists were a bit happier. We implemented a few other morale-boosting ideas, but before we could do anything substantial, Max and I were both called to higher ground.

In March 2007, Stern asked me to go to NASA Headquarters in Washington, D.C., to serve as his senior adviser for research and analysis. Alan had recently been appointed the NASA associate administrator (AA) of the Science Mission Directorate (SMD), and he selected four people (including Nobel laureate John Mather, from NASA Goddard Space Flight Center) to focus on areas where he hoped to make improvements in the efficiency of SMD operations. My area was oversight of the research program efficiencies in all four science divisions (Astrophysics, Planetary, Heliophysics, and Earth Sciences). The next twelve months were the most challenging in my career, but in many ways they were also some of the most satisfying. Stern is very clever, driven, and disciplined. He works hard, but more than that, he works efficiently and accomplishes more in any given day than you would think possible. He expected, and enabled, our team to meet milestones that some would have called miracles, and at the end of the first year of his tenure as AA for SMD, we had covered more than two-thirds of what he had outlined for our first *two* years. After my first few weeks at Headquarters, I knew I needed help, so I once again called on Max Bernstein to serve as my deputy. He moved to Washington, D.C., and together we addressed everything and more on Alan's list for improvements in the research programs. My husband, Dale Cruikshank, had moved to the East Coast with me and conducted his NASA work remotely, at the nearby Carnegie Institution, during the time we were in Washington, D.C. We all made serious sacrifices to meet the opportunity created when Alan was selected to be the next AA for SMD at NASA. Speaking for myself, I found the chance to step into the machinations of NASA at that high a level, under the leadership of someone like Alan, to be an amazing experience. Dale and I thought we had relocated to Washington, D.C., for the long haul, but in March 2008, for good reasons that are his story to tell, Alan Stern resigned from the SMD position. He asked me to stay on at NASA Headquarters, which I did for three more months, but it was not the same once my "champion for change" had left the building. I requested a change of duty station to NASA Ames and was warmly welcomed back, but still not as the researcher I had once been. I was promoted.

The NASA Ames Center director in 2008, Pete Worden, promoted me to a newly created position of "number four" in Ames Center management. I was now the deputy associate director of NASA Ames Research Center. I had oversight for many of the activities Ames was starting up, and it was an interesting time in the development of private-public partnerships. I remained in that position for two years, and having seen life from this level, I knew I did not want to go any higher in center management. I really wanted to return to my scientific research, but Pete found the next best leadership position for me instead, appointing me to be the director of the NASA Lunar Science Institute (NLSI) in July 2010. Stern had created the NLSI in 2008, and David Morrison had been the first interim director. This was a virtual institute focused on the Moon, with researchers across the country and international collaborations across the globe. I was honored to serve in this capacity and help grow the lunar science community (which has declined after the Apollo years), and I was finally thinking about research problems again, even if they were not my own. In 2013, NASA Headquarters decided to expand our purview, and I became the director of a new virtual institute that incorporated the NLSI into it. The new institute was named the Solar System Exploration Research Virtual Institute (SSERVI) (figure 9.3), and it expanded the study of the Moon to include studies of near-Earth asteroids and the moons of Mars, as well as integrating applied research (exploration) with basic (scientific)

FIGURE 9.3 Yvonne Pendleton in 2013, director of NASA's Solar System Exploration Research Virtual Institute (SSERVI). SSERVI was the successor institute to the NASA Lunar Science Institute (NLSI). Yvonne was the director of the NLSI and SSERVI, at NASA Ames Research Center, from 2010 to 2018. Courtesy of Teague Soderman.

research.[8] The success of the institute was largely due to the support we had from Jim Green, who was then the head of the Planetary Science Division at NASA Headquarters. Without Jim's strong belief in what we were doing, I am sure we could not have survived the political winds we faced, and I am very grateful for all he did.

The NLSI/SSERVI years were wonderful in terms of both the people with whom I worked and the exciting results our teams discovered. I met the remaining Apollo astronauts and heard their stories. In many ways, this felt like coming full circle in my quest from that Key West backyard vision of rockets overhead. I have been generously recognized for my leadership abilities during my "management years" by receiving several awards, including the Business Journal award for one of the most influential women in Silicon Valley (2015), the NASA Outstanding Leadership Medal (2016), and the Presidential Rank Award for Meritorious Executive (2017). The excitement was building for the launch of the JWST, however, and responding as if to a siren's call, I just *had* to go back to my own research. Lori Glaze, current head of the Planetary Science Division at NASA Headquarters, generously supported my transition back into research (financially), honoring a com-

---

8. Timothy Glotch, Gregory Schmidt, and Yvonne Pendleton, "Introduction to Science and Exploration of the Moon, Near-Earth Asteroids, and Moons of Mars," *Journal of Geophysical Research: Planets* 124 (2019): 1635–38.

mitment that her predecessor, Jim Green, had made before he became the NASA chief scientist; and the current Ames Center director, Eugene Tu, graciously allowed me to return to the NASA Ames Space Science and Astrobiology Division. I owe an immeasurable debt of gratitude to everyone I have named here, because I would not be the scientist I am today, nor would I have appreciated what I now get to do, had I not experienced the hard work it takes to actually lead these endeavors. The institute I worked so hard to build was left in the very capable hands of my former NLSI/SSERVI deputy and lifelong friend, Greg Schmidt, and it continues to thrive.

The Grand Finale

I stepped down as director of SSERVI in October 2018 and returned to my starting point in the Space Science and Astrobiology Division at NASA Ames Research Center, although now as a senior scientist. I began a deep dive back into the literature to come up to speed on many different fields in order to write compelling observing proposals for the JWST. The result

FIGURE 9.4 Photo of Yvonne Pendleton standing in front of the James Webb Space Telescope display at the NASA Goddard Space Flight Center in 2018. Courtesy of Stephanie Milam.

was that nine of the twelve proposals I worked on, with teams around the world, made it through the highly competitive process, and we were awarded significant amounts of JWST observing time in Cycle 1 and in the Early Release Science phase (more than 250 hours in total) (figure 9.4). Our proposals were structured to obtain critical data (new puzzle pieces) on dust evolution, from the diffuse interstellar medium, through various stages of star formation, to protoplanetary disks and the most primitive objects in our solar system. The anticipated new discoveries will draw astrophysics and planetary science even closer and should allow us to trace the origin of organic molecules from ice and dust in interstellar space to dust around newly forming stars and to planetary systems, including our home in the solar system.

My research story doesn't end here, because of the many students, postdoctoral fellows, and other colleagues with whom I have had the good fortune to work and are too numerous to name, but especially because of my three closest, currently rising stars: Laurie Chu, Danna Qasim, and Giulia Perotti. The wonderful thing about a career in science is that you can remain useful as long as you still have the impetus to learn more. I grew up in the age of Apollo, but I live happily somewhere between the present and the future.

Thank you, NASA, for my dreams and my reality!

The Moon and Me

was a teenager during the fifties, when the fledgling space race filled newspapers, magazines, and radio news. I was enthralled because I already
knew from science fiction stories, movies, and the *X Minus One* radio
series that technology developments were propelling us into a wondrous
future. I knew that science would transform fiction into jobs and exploration
that would fill my life.

In the early fifties, when my family lived in Miami, I discovered a series
of sci-fi books whose inside covers showed a montage of drawings of spaceships, planets, and aliens—I read them all. On a warm night, I witnessed my
first eclipse of the Moon. It transfixed me—how could a shadow be so huge
and yet invisible until the Moon passed through it? In 1956, a neighbor let
me look through his telescope at Mars, a ruddy disk with a bright spot on
one edge that I learned was the icy south pole. I was looking at another world
millions of miles away, the one that was the center of so many stories I had
read. Before Carl Sagan became known, I was hooked.

At some point my parents got me a collapsing spyglass—the first of my
twelve telescopes. In 1955, I upgraded to a 3-inch Edmund Newtonian telescope (twenty-nine dollars), using money made setting up summer carnival
rides. A primitive scope by any standard but sufficient to satisfy my urge to
observe.

In September 1956, I started at La Puente High School, twenty miles east
of Los Angeles. I was quiet, not in band nor sports. In sophomore year, Floyd
Herbert came to the school, becoming a lifelong friend. He was probably a

genius, also liked astronomy, and a long time later was the best man at my wedding. In the high-school library, I discovered another lifelong companion, *Sky & Telescope* (*S&T*) magazine. The first issue I read had Einstein on the cover—he had just died. And in 1957, when I watched Sputnik cross the dark sky, it was alone in orbit, but from *S&T* I knew more rockets were coming, and we would go to the Moon. Could I possibly be one who does it?

Floyd and I and three other students interested in astronomy started sneaking onto Mount Wilson and Palomar Observatory grounds to spend nights observing with our tiny scopes next to giant ones. One night at Mt. Wilson, a thick cloud deck covered Los Angeles, which is five thousand feet lower in elevation. With city lights hidden, stars were everywhere; so many that constellations were hard to identify. The Milky Way was visible, and the night sky was a star-studded dome above our heads, just as ancients saw it. For my senior year science fair, I built a Newtonian telescope with a 5-inch plate glass mirror. I ground, polished, and silvered the mirror, as well as building a square wooden telescope tube, with oak tripod and water-pipe mount. My physics teacher awarded a grade of A as soon as he saw the tripod. I kept the scope twenty years, finally selling it to a colleague at Johnson Space Center.

In August 1960, Floyd went to Caltech, and I left for the University of Arizona in Tucson. Naïvely, I chose UA because it had the largest telescope, 36 inches, on any campus, but unbeknownst to me, UA was about to become the most important place in the world for planetary science—my interest. When I arrived at UA, majoring in astronomy, I explored a campus of red-brick buildings, with fourteen thousand students, and out-of-state tuition of $250 per semester. I received a National Defense Student Loan, thanks to the government's panicked reaction to Sputnik. The United States needed more scientists and engineers, and I wanted to be one.

Steward Observatory was an octagonal white building topped by the dome of the 36-inch telescope. As a student, I got limited access to that telescope by helping visitors observe after monthly public astronomy talks (figure 10.1). I was allowed direct access to a 4.3-inch Clark refractor and learned to observe, drawing lunar craters, Jupiter, Saturn, and comets.

Perhaps the world's leading lunar and planetary scientist, Gerard Peter Kuiper, often referred to simply as GPK, the former director of Yerkes Observatory, moved to UA the same time I did. I was an early hire in 1961 of Kuiper's Lunar and Planetary Lab (LPL). Bill Hartmann, Dale Cruikshank,

FIGURE 10.1 Chuck Wood at the eyepiece of the 36-inch reflector of the University of Arizona's Steward Observatory, with Ewen Whitaker behind Wood's head and Alika K. Herring behind Wood's waist, 1963. Courtesy of the collection of Charles A. Wood.

and Alan Binder also came about that time—all new LPL grad students three to four years older than I was. I was at LPL for four years, working with D. W. G. (Dai) Arthur, a Welsh lunar cartographer, and Ewen Whitaker, who had started out at the Royal Observatory, Greenwich, working on UV spectra of stars before falling in love with the Moon. I measured lunar crater diameters, sharpness, positions, and morphology. I checked all work done by other measuring assistants and became very familiar with the lunar surface. This was beginning preparation for Apollo lunar exploration, for my lifetime of scientific fascination with the Moon, and a wonderful opportunity for an undergraduate. Our *System of Lunar Craters* catalog and map became the official International Astronomical Union (IAU) nomenclature, and I was a co-author. Sadly, it was published the day President Kennedy was assassinated. I walked in a daze. At the Lunar Lab, everyone watched TV and sobbed.

In the sixties, Kuiper worried that student unrest would cause a takeover of LPL because of our federal funding; LPL was disparagingly called the Dollar and Monetary Lab by other university scientists jealous of our federal funding. Because of my long hair, Kuiper thought of me as a radical student insider and called me to his office on Saturday mornings to ask about threats. There were none. But GPK wanted to talk and told me stories of capturing von Braun and other German rocketeers before Soviets could get them at the end of World War II. And he reminisced that after getting his PhD in 1933 in Holland, he had planned to go to a Dutch-run observatory in Java. He didn't, but his classmate who did was killed by the Japanese in the late thirties.

Kuiper was the lead scientist for NASA's Ranger program, America's first mission to the Moon. The Ranger spacecraft were designed to crash onto the lunar surface, taking increasingly high-resolution photographs all the way down. Rangers 1 through 6 failed, but Ranger 7 spectacularly succeeded, taking extreme close-ups of the lunar surface. On national TV, Kuiper said, "This is a great day for science and a great day for America." We were all patriotic and proud, and we immediately started measuring and counting Ranger craters. In February 1965, Ranger 8 crashed—as planned—into Mare Tranquillitatis (Sea of Tranquility), taking pictures all the way down. Alika Herring and I used the Kitt Peak 84-inch telescope to look for an impact-generated dust cloud. We saw no cloud, but in moments of steady air, we detected clusters of small secondary craters that had probably never been seen before.

I graduated with a BS in astronomy in 1965 and continued another year at LPL, but I had no plans or dreams; I needed to do something different. I applied to be a Peace Corps volunteer—a tie to JFK, my hero. A couple of weeks later, I received a telegram saying I had been accepted and assigned to Kenya. I was excited and got out an atlas to see where Kenya was. It's in East Africa, where many wild animals still roamed, and volcanoes were everywhere. I happily accepted the offer. Bill Hartmann said I was courageous to go. Kuiper gave me a going away party with a Chinese takeout meal at his home. I was honored and didn't know why he did it, but I never saw him again, for he died while I was in Africa.

After our long flight from New York via Amsterdam, Cairo, and Entebbe, a planeload of tired Peace Corps volunteers arrived in Nairobi at 2 a.m. Sigalame, my school, was far upcountry, on the border with Uganda. I taught math, physics, a few lectures of biology, and religious knowledge to classrooms of fourteen- to eighteen-year-old boys (figure 10.2). Kenya is bisected by the equator, a very special place on Earth, where all the stars in both hemispheres of the sky are visible. In 1967, I created a star chart for the equator; the Peace Corps mimeographed it and sent it to all schools in Kenya. There was no evidence that it was ever used, but *S&T* published a mention of it.

Frequently, I traveled to explore the volcanoes of the great East African Rift Valley, which slices through Kenya. Five years later I used this Kenya volcano experience and lava tube photos in an article Dale Cruikshank and I published about the origin of rilles—narrow sinuous channels on the Moon.[1] A later paper compared possible rift valleys on Mars, Venus, and Earth, based partly on my explorations of Ethiopia and Kenya rifts.[2] Don't just look at scenery, understand it.

FIGURE 10.2 Kenyan students using a small telescope. Courtesy of Charles A. Wood.

After two years of Peace Corps service in Kenya, I took a boat down the Nile to Cairo, the beginning of an eight-month trip home. I hitchhiked through the Middle East, north through Eastern Europe to Soviet Russia, and back west and south across Europe. I visited the National Observatory of Athens to see the telescope famed astronomer Julius Schmidt used to map the Moon in the 1870s; I also enjoyed tasty gyros from street vendors. In Bavaria, I visited my LPL friend Alan Binder, who took me to a church in Nördlingen, not for its architecture or history. The stones it is made of were melted fifteen million years ago by an asteroid impact, which excavated the fifteen-mile-wide Ries crater that the cosmic church sits in. Are the churchgoers closer to God because their church's stones came from the heavens?

Florence was next, with the Galileo Museum and its priceless items— Galileo's telescopes, Moon drawings, and middle finger. Near Rome, I visited the Papal Palace of Castel Gandolfo, the traditional summer palace of popes

---

1. D. P. Cruikshank and C. A. Wood, "Lunar Rilles and Hawaiian Volcanic Features: Possible Analogues," *Moon* 3 (1972): 412–47.

2. C. A. Wood and J. W. Head, "Rift Valleys on Earth, Mars and Venus," in *Tectonics and Geophysics of Continental Rifts*, edited by Ivar B. Ramberg and Else-Ragnhild Neumann (Dordrecht: Springer, 1978).

FIGURE 10.3 Decades after the author's post–Peace Corps visit, he returned to Castel Gandolfo with his wife, and Vatican astronomer Brother Guy Consolmagno showed them ancient telescopes and original copies of Newton's *Principia*, as well as many classic Moon books and maps from the 1600s. Courtesy of the collection of Charles A. Wood.

and headquarters of the Vatican Observatory; it is built on a volcanic crater (figure 10.3).

During the southern Italy section of my European travels, I climbed Vesuvius and visited its namesake volcano observatory in Naples, built on volcanic rocks that erupted in the 1500s. (Frequent earthquakes and rising ground suggested another eruption was imminent.) In the Aeolian Islands, I climbed Stromboli volcano alone (which was rather stupid since there would have been no help had I been injured). Boulders ejected from the summit vent followed parabolic arcs, rising up and then falling to the ground. I would look up to see where they were headed and then step out of the way. One of my pictures later became the frontispiece for *Basaltic Volcanism on the Terrestrial Planets*, which I co-edited. In Sicily, I climbed Etna volcano, which had erupted recently. Layers of snow from the previous winter were protected by overlying ash layers—a hot and cold sandwich; I wonder if there are any on Mars.

I returned to UA in 1969 to start an MS degree in geophysics and resumed work at the Lunar Lab. I started a new lunar crater catalog, measuring diameters and depths on Lunar Orbiter 4 spacecraft images, with ten times better resolution than prints used for our previous crater catalog. It was the best lunar catalog—but never published. In the midsixties, Bill Hartmann and Kuiper had discovered impact craters so large that Bill had named them lunar basins. In 1971—fifty-plus years ago—Bill and I used Lunar Orbiter 4 images of the entire Moon to discover more basins on the farside, and we published descriptions and an age sequence of all the Moon's basins—a fundamental paper, still cited.[3]

My MS thesis was a geophysical study of the mile-wide MacDougal crater in the Pinacate volcanic field, just over the Arizona border in Mexico. Using a handheld magnetometer, I conducted a magnetic survey, walking back and forth across the crater floor, up the walls, over the rim, and out hundreds of feet onto the surrounds. For comparisons, I conducted magnetic surveys across other nearby maar craters and a gravity survey (much harder—especially in the August heat) of MacDougal. My adviser also flew a magnetometer survey of the entire volcanic field, with me lying at an open hatch in the floor of the small plane to photograph where we flew. I published a paper derived from my thesis work in *Bulletin Volcanologique*, the classic journal for volcanic studies.[4]

In March 1970 LPL scientist, Carmelite priest, and pilot Father Godfrey Sills flew Dale, Floyd, and me to witness the solar eclipse in southern Mexico. Our single-engine plane was not allowed to depart before 6 a.m., but airspace closed at six for the eclipse. We barreled down the runway exactly at six, and flew to a short dirt runway in the middle of the eclipse path that we had found the day before. We buzzed the strip to scare off cows and boys, and then came in for a dusty landing. Just before the eclipse, another plane landed, causing confusion and dust. The Moon's shadow came from the west, darkened the sky, and swooped over us for three minutes of totality. As the Moon fully covered the Sun, everyone "aahed" and cameras clicked when the

3. W. K. Hartmann and C. A. Wood, "Moon: Origin and Evolution of Multi-ring Basins," *Moon* 3 (1971): 3–78.

4. C. A. Wood, "Reconnaissance Geophysics and Geology of the Pinacate Craters, Sonora, Mexico," *Bulletin Volcanologique* 38 (1974): 149–72.

diamond ring popped out; I looked with my eyes to see sunset in 360 degrees and beautiful coronal rays.

In January 1973, MIT professor Tom McGetchin invited me on a trip with his students to erupting Pacaya volcano in Guatemala. I saw my third eruption, with thick, sluggish 'a'ā lava flowing downslope—beautiful at night, much slower than fluid Hawai'i flows. One night, sitting around the campfire, I mentioned to Tom that a paper he had sent me prior to publishing it appeared to have an error. His correction for air resistance for ejecta flying through air made it fly faster—it should have slowed down. Instead of being irked to have this error pointed out to him, Tom was appreciative. Later, this seemingly insignificant incident would change my life.

As I was finishing my MS degree in geophysics, I saw an ad for a two-year position at a university in Ethiopia, which made Kenya look tame and well explored in comparison. Ethiopia had been an independent nation for 1,500 years, with the Queen of Sheba, King Solomon, and volcanoes everywhere; I applied. In 1973, the Geophysical Observatory (GO) at Haile Selassie I University in Addis Ababa, Ethiopia, hired me. En route, I saw my second solar eclipse, this one from the summit of Marsabit volcano in northern Kenya. Being on a mountain in the middle of a desert gave a panoramic view of the racing Earth shadow that passed overhead as the eclipse ran its course.

Canadian Jesuit priest Pierre Gouin started the GO during the International Geophysical Year (1957–58). He, short-term faculty like me, and Ethiopians kept the GO going, surviving drought, earthquakes, and a revolution. One week a month I was responsible for keeping the observatory running. Every day I held seismometer paper over a flame to coat the paper with carbon, wrapped it around a seismometer drum, read the previous day's seismograms (scratched through the carbon), and reported any activity, plus magnetometer and meteorology readings, to international geophysical bureaus.

The other three weeks, I was free to carry out geophysical research, such as gravity and magnetic surveys of calderas in the Ethiopian Great Rift Valley. Frances Dakin and other geologists at Haile Selassie University and I did fieldwork in the Rift Valley and Afar—a triangle of land below sea level kept dry by lava flows, which blocked a trough that previously let in the Red Sea (figure 10.4). Afar is the only place on Earth where you can walk on dry seafloor. Two ocean-spreading centers (the Red Sea and the Gulf of Aden) and the East African Rift Valley meet in Afar. Ultimately, the opening of the Rift Valley will cause Somalia, eastern Ethiopia, and Kenya to split from Africa, letting ocean water separate the continent and a sliver of East Africa that will move away, just as Madagascar did 180 million years ago.

During January 1974, we made the arduous and dangerous trip to the Afar Depression and Erta Ale volcano. Erta Ale continuously erupts within the lava lake in its summit caldera. The thin crust is surrounded by cracks where red lava oozes out, pushing crust pieces apart—plate tectonics at a scale of a few meters. At night, the lava was eerie red, and expanding gas propelled magma upward, forming large red-orange bubbles that popped as their skin was torn

FIGURE 10.4 The author measuring the gravity field inside an Ethiopian caldera, with Frances Dakin (*right*) recording readings and kids wondering what they're doing. Courtesy of the collection of Charles A. Wood.

apart. Ropes of lava fell back into the moltenness, frighteningly mesmerizing. We turned our backs on the lava's redness to observe comet Kohoutek in an absolute black sky, with no city lights for one thousand miles to the north.

A colleague and I published in *Nature* a paper—my first in that prestigious international journal—suggesting that sunspot cycles had influenced droughts over more than one thousand years of Ethiopian history.[5] We then examined rainfall data for all of Ethiopia. The strong correlation between minimum solar activity and droughts meant we could predict and prepare for the next drought by finding the date of the next sunspot minimum. But sunspots don't behave. Spot cycles range from seven to fifteen years, and we needed accurate prediction of when the next minimum would occur. I analyzed data, discovering that the number of spots at sunspot minimum is related to the time of the next minimum; I predicted that the likely minimum of the coming cycle was early eighties. The next big drought occurred in 1983–85.

In September 1974, a coup d'etat replaced Emperor Haile Selassie with the Derg military junta. Our GO seismometers detected troop movement. And our GO dog, Milligal, was killed by troops marauding through campus.

---

5. C. A. Wood and R. B. Lovett, "Rainfall, Drought and the Solar Cycle," *Nature* 251 (1974): 594–96.

I almost was; I came face to face with a nervous young soldier who pointed his rifle at me. He turned and ran; I didn't die. Two years later, my Geology Department colleague Bill Morton was driving a Land Rover to help three enemies of the Derg escape. Bill and the others were caught and murdered on the side of the road.

As I got ready to leave Ethiopia, I wrote Tom McGetchin, leader of the Guatemala volcano trip I had joined, about a possible job with him. As noted earlier, I had found an error in one of his manuscripts. Realizing from that just how sharp I was, he generously recommended that instead of working for him, I ought to go to grad school to earn a PhD. He told Jim Head at Brown University about me, and he agreed that I should apply. My application was successful, and I received research fellowships that paid all expenses and salary for three years. All as an indirect result of my finding that error in Tom's manuscript.

The planetary science group at Brown University was one of most active in world, under the leadership of Thomas A. (Tim) Mutch and Jim Head. In the seventies Tim produced authoritative books about the geology of the Moon and Mars and went on to serve as leader of the Viking Mars lander teams. (He was tragically killed in 1980 in a climbing accident in the Himalayas. In his honor, the Viking 1 lander, which was still sending images and data back to Earth at the time of his death, was formally renamed Thomas A. Mutch Memorial Station.) Jim had trained Apollo astronauts and was and still is a leading lunar scholar. Wow! For the next few years I worked with my advisers and fellow grad students; this was the first time I was an integral part of a research team.

I soon got into Head's routine of submitting abstracts to five or six conferences a year and publishing a steady stream of papers. It was the most productive time of my career, with ten to twelve publications a year, all co-authored with Jim, Tim, and fellow students. Many of my papers analyzed the number, distribution, and nature of impact basins on the Moon, Mars, and Mercury. Others explored the rates and morphological patterns of growth of Earth's cinder cones and stratovolcanoes. These papers are still commonly referenced forty years later.

In August 1978, I defended my PhD dissertation and left for Washington, D.C., the next day. I started a one-year postdoc at the Smithsonian Institution's Natural History Museum, working with Tom Simkin on volcano growth patterns. On my first day of work, Tom asked, "There has been an eruption in the Galapagos Islands; do you want to go?" Answer: Yes! We used a small boat to motor around volcanic islands to reach and climb Fernandina volcano, which is a model for much larger Olympus Mons and other flat-topped Mars volcanoes.

For 1979–80 I was awarded a senior postdoc position at NASA Goddard Space Flight Center in Maryland, studying rift valleys. I compiled satellite images, geology, and gravity data to understand how rift valleys form. (It would have been easier after 1987 when Photoshop was invented.) Living near rifts in Kenya and Ethiopia gave me personal familiarity with their long linear valleys, stairstep scarps to the bottom, and floor volcanoes strung out like beads along a necklace. I compared Earth rifts with ones on Mars and Venus, as well as co-authoring a paper about an explosive ocean-floor volcano—the water pressure is nearly the same as atmospheric pressure on Venus.[6]

While finishing up at Goddard, I applied to Johnson Space Center (JSC) in Houston. Jim Head had recommended me. I was duly hired and worked in the Space Science Division, where lunar samples are housed and studied. Many of the scientists had trained Apollo astronauts and still studied lunar samples. After I had been there about six months, I asked my supervisor what I was supposed to do. He said, whatever I want. Think and write. Although I had previously studied the Moon, Mars, and Venus, at JSC much of my work was about meteorites. I compiled a catalog of ordinary chondrites, leftover pieces of the rocks that formed the planets 4.5 billion years ago. *Meteorites* journal refused to publish it, saying it would be out of date next time a meteorite fell. This is just as stupid as saying a dictionary is outdated once a new word is added. The real problem, of course, was that I was not a member of the meteorite fraternity of planetary scientists.

One day I noticed, taped on the office door of a colleague, a one-paragraph abstract of a talk he had given recently. The last line was that three kinds of meteorites are unusually young—about 1.3 billion years old—and might have come from Mars. This excited me. Perhaps we had pieces of Mars in the building! I told a colleague, Lew Ashwal, who knew far more about rocks than I do, and we started researching these meteorites. There are three related types, shergottites, nakhlites, and chassignites, named after the places where they were first found. Rather than repeatedly saying the three names,

---

6. Patrick T. Taylor, Charles A. Wood, and Timothy J. O'Hearn, "Morphological Investigations of Submarine Volcanism: Henderson Seamount," *Geology* 8, no. 8 (1980): 390–95.

I called them SNCs, or Snicks, and though at first some of the meteoriticists complained, the name appears to have stuck. Lew and I found that the SNCs had different chemistries, ages, and isotopes compared to all other meteorites, which come from asteroids orbiting the Sun in the space between Mars and Jupiter. So, where were SNCs from?

Unlike most meteorites, SNCs are igneous rocks—they melted inside a larger body. They are very young, with chemical compositions similar to that of the Martian surface, isotopes unlike other meteorites, and in some, water. Lew and I concluded that Mars was their most likely source. Talks for Lunar and Planetary Science Conferences (LPSCs) are accepted and arranged into a program by a committee. In 1981 I was on the committee and overheard a leading meteoriticist, who didn't know me, say that Lew's and my abstract was rubbish and couldn't be in their session; they dumped it instead into the Mars session. When the title on the talk was announced at the 1981 LPSC, some people snickered as I walked to the lectern. When I finished reporting the evidence Lew and I had found, and our theory that the meteorites are from Mars, there was disbelief. As questions ended, I pointed out that SNCs may have the same origin as Snickers candy bars—How so? Snickers are made by the Mars Candy Company.

"When you have eliminated the impossible, whatever remains, however improbable, must be the truth." Lew and I placed this well-known Sherlock Holmes quotation (from *The Sign of the Four*) at the beginning of the article we published, "Meteorites from Mars," in 1981. Forty years later, 266 SNCs have been found, and the meteorites from Mars hypothesis is widely accepted.

JSC is home of the astronaut corps. Most were young astros waiting for a ride on the new space shuttle. After the first shuttle mission, friends and I went to our favorite seedy Mexican café, and there were John Young and Bob Crippen, astronauts who had just landed the shuttle the day before. We said hello. While Young and Crippen had been in orbit, I had learned that Mt. Etna in Sicily was erupting. If I could have gotten a message to them, they could have photographed the eruption from space. One space image can show the entire area affected by an eruption, making a more accurate map faster than volcanologists on the ground could. Before the next space shuttle flight, I discovered JSC's Space Shuttle Earth Observations Office (SSEOO) and was invited to participate in sending messages to orbiting crews and planning what targets astronauts should image. Within six months, the SSEOO leader retired, and I replaced him. For the next five years, our team of fourteen trained all the astronauts who flew on the shuttle in techniques for imaging the Earth. I also worked with a Canadian impact crater expert, Richard Grieve, who spent a year in Houston writing *Astronaut's Guide to Terrestrial Impact Craters*, which NASA's Lunar and Planetary Institute published in 1988. Our goal was to train astronauts to detect unknown craters from orbit. They did and still do.

NASA would sometimes fly civilians for political or educational reasons. The most famous was when the Challenger space shuttle launched on January 26, 1986, with Christa McAuliffe, a teacher, on board (figure 10.5). Everyone knows that a Challenger solid rocket exploded, killing the astronauts. I

FIGURE 10.5 Christa McAuliffe in foreground, with backup teacher Barbara Morgan, Wood, and coworkers at the Johnson Space Center, studying astronaut photos on a light table. Courtesy of Charles A. Wood.

had helped Christa prepare a lesson called "Earth from Space," which she had planned to teach from orbit. I helped select photos and described what they illustrated, in case the sky was cloudy during her mission. The day before launch, she called me from the cape to review her lesson. I told her that she had it down and would do a great job.

I watched the Challenger launch from our building's conference room with twenty other NASA scientists. As soon as the rocket exploded, we all realized what had happened, and disbelief and sorrow saddened every face. At that moment, I gave up my own hope of someday flying on a shuttle mission.

In 1990, astronaut Jim Buchli told me that a chair position for a Space Studies Department was open at the University of North Dakota (UND). I guffawed at the idea of living there but interviewed for the position, just to visit a state I'd never been to. It was a small department with five enthusiastic faculty. Space Studies combined science, engineering, social issues, and law into an interdisciplinary master of science degree. Seventy percent of the students were U.S. Air Force officers at Grand Forks and Minot Air Force bases. A faculty member went to these bases for five-hour classes each Monday night. I accepted the job offer, and our family arrived in North Dakota in July 1990. (The snow had just melted!) During the same year, Cambridge University Press published my first major book, *Volcanoes of North America*, which describes 254 young volcanoes in the United States and Canada. It was co-edited with Jürgen Kienle, who organized the Alaska section.

In 1990 I was selected to be a member of the radar team for NASA's Cassini mission to Saturn. The spacecraft launched in 1997, finally arriving at Saturn in 2004. Saturn's moon Titan is covered in thick clouds, so its surface was excitedly unknown—our radar team would be the first humans to see it. Titan is bigger than planet Mercury, and Cassini data showed that it is the only world other than Earth with lakes, seas, and rivers. But the liquids are methane and ethane at a temperature of −290°F, immensely too cold for liquid water. The crust of Titan is water ice, veneered with organic materials falling out of the atmosphere. Deep under the ice crust is a water ocean. Over the twenty-seven years of Cassini, I worked on many projects, including cataloging Titan's impact craters, as I had done fifty years earlier for the Moon. I found only about fifty craters. This means that Titan's surface must be very young, about 500 million years, which is also, in fact, the average age of Earth's surface. Since many more craters than that should have formed, clearly some kind of geological processes must have removed them.

Also in the nineties, I was awarded a $1 million competitive NASA grant. We started VolcanoWorld (VW), an educational website with Rocky, the volcano creature, and hundreds of pictures and much information about volcanoes. Four million kids and adults visited VW each year. In 1995, a nationwide military base closing took away many of our students from local air force bases. I decided to convert our base-visiting version of space studies to Space.Edu, taught live over the internet, to continue serving our air force students. I presented the idea for an online space studies MS program to university officials. The UND lawyer said that we should not be the first to do this, since it wasn't clear what the legal issues might be. I replied that if humans followed advice like that, we'd still be living in caves. The overwhelming vote was to start Space.Edu.

While at UND, my book *The Modern Moon: A Personal View* was published by Sky Publishing Corporation. During the previous eighteen months, when the book was reviewed, edited, and designed, *Sky & Telescope* began publishing monthly two-page articles excerpted from the book; twenty-three years later, my lunar articles are still appearing in my favorite magazine.

In January 1993, as part of a NASA-funded space-radar study of Northern Andes volcanoes, I traveled to Ecuador to study the volcanoes' spacing. Other researchers and I first drove sixty miles north to Pasto, Colombia, for a ninety-person United Nations (UN) scientific workshop on volcano safety. On Wednesday, January 14, jeeps carried some of us to the summit of Galeras stratovolcano, west of Pasto. At fourteen thousand feet, the summit was cold; fog came and went, revealing and hiding a steaming cone on the caldera floor.

After looking from the rim, we divided into two groups. My group examined ash draping the outer rim of the caldera, while the other hiked into the caldera to sample gases from the cone. We drove along a rugged trail, stopping to see ash deposits from past eruptions. At our lunch stop, we felt an earthquake, heard a BOOM, and saw ash shoot above the caldera. An eruption had occurred, and no one could say if another was coming. Our two local volcanologist guides, Marta Calvache and Patty Hall, told us to drive off the mountain as fast as possible. The two women drove quickly back to the summit, finding refrigerator-size and smaller boulders everywhere. They courageously went into the caldera, shouting to locate survivors. They found Stan Williams, the conference leader, with broken legs and boulder wounds on his back, head, and arms. They carried him out on a stretcher, saving his life. Searchers later found six dead scientists and three tourists. If I had chosen to go inside the caldera, likely I would not be writing this tale.

On January 1, 2000, I started the website Lunar Photo of the Day (LPOD). I posted a daily image of the Moon with a short text for eleven years, and now LPOD is recycled day by day (https://www2.lpod.org/). A few years later, I started the Moon Wiki, which features a page for each named lunar feature and has discussions of maps, nomenclature, and the history of observations (https://the-moon.us/).

After ten North Dakota winters, I applied to be director of education at Biosphere 2, forty miles north of sunny Tucson, Arizona. Bio2 is the world's largest (three acres under glass) test site for experimenting with future climates. They hired me, and our family drove south in the winter. With no interaction with Bio1—Earth—except sunlight, electricity, and phone calls with Bio2's HQ, eight Biospherians (Bios) lived in a sealed glass dome, growing their own food and recycling air, water, and waste. Their vision—hippieish, almost quasi-religious, and based on the ideas of the guru-like British independent scientist, environmentalist, and futurist James E. Lovelock—was that, like an aquarium, Bio2 would stabilize and sustain them. The entire Earth, including rocks, plants, and air, is a living organism that maintains an environmental balance of temperature, oxygen, and oceans so that life can continue to exist. But Bio2 didn't stabilize, and the Bios cheated to make it seem to. They lost all credibility.

After that, no university would touch Bio2. Finally, Mike Crow, then executive vice provost of Columbia University (and now president of Arizona State University) accepted the challenge, and repositioned Bio2 as a climate research lab, with no one living inside. Crow also started a semester abroad–type program, where undergraduates from Columbia and other universities came to the Bio2 desert facility for a wildly interdisciplinary environmental science and policy experience, centered on field trips across Arizona and into Mexico. Our Bio2 program included art as well as science and policy. Students wrote and performed plays, strummed guitars, and created artwork inspired by what they experienced. One of my favorite class reports was a tap dance performance illustrating the plate tectonic opening of the Gulf of California. Students absolutely loved Bio2, but a new president at Columbia University ended it, and I was, for the only time in my life, fired.

Being in my early sixties, I saw little chance for finding a new job, so in November 2004, my wife, Vera, and I visited Nicaragua as a possible retirement locale. I talked with the Nicaragua Geologic Survey about a possible job monitoring volcanoes. In January 2005, we moved to the lovely land of volcanoes, good food, and cheap living. I left six months later; retirement failed.

In May 2005, I was hired as director of the Center for Educational Technologies (CET) at Wheeling Jesuit University in West Virginia. The research center for the NASA Education Office, CET had started in 1994 with fanfare and earmarks to construct a building. CET received funding over the first years I was there for three projects exploring newish technologies and the Moon.

- We developed BLiSS, an iPad app that simulates the growth of plants hydroponically. Which plants would maximize production of food and oxygen to support six astronauts at a lunar base? Hope you like lettuce.
- MoonWorld: Exploring the Moon in virtual worlds was our Second Life internet project, in which students built space-suited avatars to explore (by walking and driving a rover) the geology and formation sequence of the Moon's Timocharis impact crater and surrounding lava flows.
- Our National Science Foundation–funded CyGaMEs Selene: A Lunar Construction GaME required students to build the Moon by accretion of debris from a large collision with Earth 4.5 billion years ago. Collisions have to be slow enough for debris to stick and grow. Many craters formed over time, and lavas covered some old craters.

In 2010, I was invited to become chair of the Task Group for Lunar Nomenclature, which is a section of the IAU's Working Group for Planetary System Nomenclature. WGPSN approves names proposed for landforms on moons and planets, a surprisingly contentious subject; I have been sworn at by a planetary scientist who didn't like one of our decisions. The names are of famous philosophers and scientists from ancient Greek times until today. You have to be dead for three years to be honored, so there is little rush, but tens of my prior friends and acquaintances now have craters on the Moon.

In 2007, Japan launched their Kaguya spacecraft to orbit the Moon. They used my 2004 Lunar 100 list of interesting places on the lunar nearside to select targets for high-definition TV imaging for public relations. I learned this when I was contacted for a similar list to guide imaging of the farside. I made a list of fifty intriguing farside features, and after the mission ended,

Moto Shirao, who had contacted me for the imaging targets, invited me to co-author the book *The Kaguya Lunar Atlas: The Moon in High Resolution*, which we published in 2011. By 2013 I had published five books on impact craters, volcanoes, and the Moon. These had been produced by major publishers, but the publishers did little to promote the books, and secondhand communication with people who did the layouts—once in India—had been difficult. I worked with an internet friend, Maurice Collins in New Zealand, and we designed, printed, advertised, and sold a few thousand copies of our *21st Century Atlas of the Moon*. It keeps being reprinted, and royalties are still arriving ten years later.

A few years ago, I was surprised when an asteroid was named for me. The citation from the IAU's Minor Planet Circulars identifies the asteroid as "(363115) Chuckwood = 2001 FW224." The citation states that Wood "made fundamental insights into the role of cratering and volcanism in shaping planetary surfaces. He tirelessly promoted science education through numerous books, popular articles, and the Internet. His development of education programs introduced many students to science."

Aw, shucks!

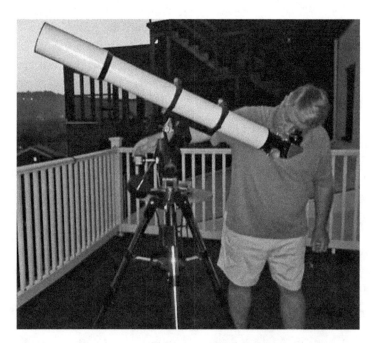

FIGURE 10.6 Observing the Moon from Wheeling, West Virginia. Courtesy of Charles A. Wood.

During the last three years, I cut down my work at Wheeling University to thirty hours per week to spend more time writing. I have published a short book, *Our Phillips House*, on the pioneer man who built our 1831 house. I greatly enjoyed research and writing about that distant time, so I also wrote a historical murder mystery novel, *Wheeling 1850*, in which an astronomy professor and his daughter study the Moon, comets, and a new star, being interrupted by horrible murders.

As for the Moon, I still use a 4-inch refractor—smaller and lighter than the wood-tubed Newtonian I built sixty years ago—from our back deck to check that my lunar crater friends are still there (figure 10.6). When I was in the Peace Corps in Kenya, I learned that a belief of my students' Luhya tribe was that when a person on Earth dies, a leaf from a lunar tree falls to the surface, accounting for the dark lunar maria. A lot of my friends' leaves have fallen, and mine will too, hopefully not too soon.

From My Backyard to Space

PAOLO TANGA

How did it all get started? Was it that book on practical astronomy, taken (without promise of return) from my uncle's library? Was it the sight of the celestial landscapes from car widows in the Po River valley in northern Italy on freezing winter nights, or just innate curiosity for nature, culture, and science, inherited from my parents? I do not know the answer, of course, but recognize in my past the accumulation of ideas and inspirations that first guided an inquisitive kid to the stars and then to a career in astronomy.

My first memory of space, if we want to call it that, was a fuzzy impression now faded with time, dating back to the night when a man first stepped on the Moon. I was just over three years old, and my parents and grandparents used to spend a few summer weeks at a hotel on the Italian Riviera. It was there I recall first asking naïve questions about nature, like "Why does the sea not drain itself like a bathtub?" Was that just my quirky sense of humor or an early indication of scientific curiosity? Who knows?

It was late on the historic night of July 20, 1969, as TV screens were bringing into houses everywhere the images of Apollo 11 on the lunar surface. The same happened in the hotel hall, where my parents decided to keep me up and involve me as history was being written. I cannot remember many details, nor my state of consciousness while waiting for those ghostly images to appear in the wee hours of that morning. Still, I have never forgotten that fuzzy black and white TV image of ghostly shapes difficult to make out, while a small crowd of people gathered around.

I have far more solid memories and fully recall my disappointment when I was seven, as the last Apollo mission was announced on TV (December 1972). That feeling certainly meant something: that I was fully aware of the thrill of seeing men on the Moon, and the captivating inspiration that this incredible and successful challenge had generated in me. No surprise, really. After all, who could have resisted the fascination of monumental rockets climbing the sky with roaring flames, bringing fragile men to the most exotic of all possible destinations? It was a place where walking and driving a lunar rover seemed so risky, funny, and outer-worldly all at the same time.

Alongside all this, my humble black and white two-channel TV set was more than enough for me to "learn" that extraterrestrial visitors posed a real menace, except for the existence of powerful secret Earth protection agencies. The British TV series *UFO*, which aired in Italy from 1970 to 1974, suddenly revealed a fantastic world to me, and that blew my mind in just a few episodes. In those same years, the Italian astronomer Paolo Maffei (1926–2009), whose name will always be remembered for the two galaxies Maffei-1 and Maffei-2, obscured by dust in the plane of the Milky Way galaxy but revealed by infrared photography, wrote his first popular book, *Beyond the Moon* (*Al di là della Luna*, 1973). Well-written and documented, it describes a hypothetical journey of exploration from the Moon, toward the planets, the stars, and eventually the distant universe. It quickly became a bestseller, which my father read with much interest. Attracted by those black and white photos of planets, stars, and galaxies, I too began to browse through its fascinating pages and then to read it fully. This book also inspired my father and me to read Steven Weinberg's famous *The First Three Minutes*, a far more challenging text, which, though not easy for me to follow at the time, provided a first taste of the role of particle physics in shaping our view of the cosmos.[1]

With my parents' rising awareness of and interest in astronomy and space, they talked of buying a small telescope, at least as a theoretical possibility. More TV documentaries on astronauts and space-related activities were also very popular, and I eagerly looked for them. My first concepts of physics go back to that time, which I am sure had a strong influence on my future choices. I particularly remember the astronauts dropping a hammer and a feather on the Moon, and the illustration, by an elastic tissue with a heavy

---

1. Weinberg, *The First Three Minutes: A Modern View of the Origin of the Universe* (London: Fontana, 1977).

ball on it, of the space-time deformation induced by gravity (but only the first part appeared logical to me). I also cannot forget that particular summer evening, in July 1975, when I watched the encounter of U.S. and Russian astronauts in space, in the so-called Apollo-Soyuz mission. I was outside on the terrace of my grandfather's house, with the TV pointed toward an open window, an ideal setup for relaxing and enjoying live space programs. That terrace also played another major role in my astronomical life, years later.

Oddly enough, however, my first optical device to explore new worlds was not a telescope but a microscope I received as a Christmas present. Its decent optical quality prompted me to investigate the intimate nature of small things: blood cells (though not mine), paramecia found in the water of flowers, fly wings, skin tissues, vegetable parts, insects—anything thin enough. I started dreaming of techniques that I was reading about, involving advanced sample staining, microtomic sections, and so on, but they seemed too sophisticated to me. In middle school, I began to experiment with a chemistry set. Of course, in addition to the mixtures suggested, I also tried some more experimental combinations, resulting in a few memorable small-scale chemical accidents. I appreciate (with some regrets) why such kits today include far fewer and less varied chemicals, as well as far more explicit safety precautions.

Despite these interludes, space remained paramount in my life. The 1976 Viking landers and their experiments looking for biological activity on Mars were featured on TV with plenty of interesting details, generating genuine excitement. Clearly, my interest was shifting more toward space exploration than the sky itself. Moreover, living in Torino, a city of a million people in northern Italy, focused on the automobile industry and with foggy skies, was not ideal to ignite the fires of astronomy. Our apartment had two large balconies, one facing west, the other east, on the top (ninth) floor of one of the tallest buildings in the neighborhood. The western landscape was more affected by other buildings and main roads, facing the Alps, but the eastern balcony offered a totally open view of a hill. Sunrises full of light, or reddened by fogs, were frequent and magical. The hill itself, inundated by warm light at sunsets, transformed at night into a sparkling swarm of scintillating distant lights—an Earthly constellation. This revelation was even more striking on windy days, when dry air flowing down over the Alps removed all fogs.[2]

---

2. This is known as the *phoen*. It is a typical meteorological configuration, in which air climbs a mountain range, releases humidity in the form of rain or snow, and descends on the other side of the mountains, completely dried up, while heating by compression.

Crystal clear stars and planets rising above the hill were there waiting for me, and the time at which I could start to name them was near. In winter, driving back from weekend visits to my grandparents in Saluzzo, my birth town, the rural sky views were full of stars. My father once stopped the car to show me the Big Dipper and Polaris, but most of the other constellations and stars remained a mystery to me. I also noticed a faint tiny dipper several years before learning about the existence of Pleaides.

One day a small astronomy handbook fell into my hands, an Italian translation of the famous *Sky Observer's Guide,* by R. Newton Mayall, Margaret Mayall, and Jerome Wyckoff, at the library of an uncle who held a master's degree in biology and had many interests.[3] His house was another of my gold mines. I confess that I never returned the book to its legitimate owner. Instead, it became my first real contact with practical astronomy, teaching me about telescopes, magnification, resolving power, the aspect of planets, star clusters, and nebulae at the eyepiece—all the basics. Plus, it portrayed the impressive backyard of an accomplished American amateur with a telescope tube so big it required a ladder to reach the eyepiece. That was my point of no return. I was filled with an insatiable desire to learn as much as I could about how a telescope worked and what it could show. The book also showed me drawings of the Moon, planets, open and globular clusters, and so on. There was not much detail, but I was introduced to many objects I now could hope to see directly and at first hand.

About this same time, I met somebody living a few blocks from me who owned a small refractor and was already participating in some meetings of local amateur astronomers. To me in the early eighties, a telescope seemed like something I could only dream of. At the time I had only some pocket money and a little savings with which I might buy a book from time to time. I was not used to managing "big" sums. Buying a telescope, scaled to that time in my life, was like getting my first car or buying my first house.

Besides, the availability of astronomical instruments was extremely limited by today's standards, at least in Italy. I remember three photography stores in Torino that exhibited some brands I was familiar with, but I had no clear idea which to get, since my main sources of information were books, some magazines advertising refractors of dubious quality, and my new amateur astronomy friends. This helped me decide what seemed to be the best

---

3. R. Newton Mayall, Margaret Mayall, and Jerome Wyckoff, *The Sky Observer's Guide: A Handbook for Amateur Astronomers* (New York: Golden Press, 1959).

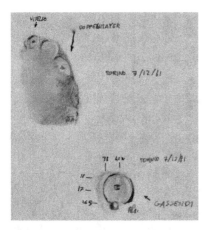

FIGURE 11.1 A large sunspot group, sketched by the author in 1981. Courtesy of Paolo Tanga.

compromise at that time, a 4.5-inch reflector, on an equatorial mount compact enough not to shock my parents, but potentially offering interesting views of the sky.

Since that was my personal decision, I also wanted to contribute money I had earned, and I spent several grueling hot days that summer painting the railing of the large terrace of the apartment, which made me fully appreciate how large that railing was. But in the end my grandfather paid me the money needed to buy the telescope I wanted. I was able to get the precious instrument in the autumn of 1981 and had my first memorable views from the eastward balcony in early October. Never having seen the Sun, the Moon, or planets through a telescope before, the views were nothing less than breathtaking. I also began to sketch some of what I was observing, and for several weeks, I followed directions in the small handbook that came with the telescope (figure 11.1). Finally, as the winter sky was gradually coming into view, Orion rose amid the scintillating landscape of the hills. In the winter, from time to time, the fog would lift over the city, revealing a dark, starry night. On milder, spring nights, the calm and beauty of the town sleeping beneath the stars evoked peaceful and deeply inspiring moments. That is how it all began for me.

Finding a Goal

The ensuing years were an extensive exploration of what amateur astronomy could offer, while always keeping an eye on space activities. I keenly followed the results obtained by the Pioneer and then the Voyager spacecrafts, and an iconic and inspiring image of Saturn and its rings graced the walls of my room. A few months after getting my telescope, I joined the national society of amateur astronomers, Unione Astrofili Italiani (UAI). At the time, the main focus of this association was to coordinate amateur research, and I quickly realized that the skies in Torino were not dark enough to invest energies anywhere else than in lunar and planetary observations. Based on information in my handbook and some other sources, coupled with the theoretical limits of my small telescope, I decided to join the Jupiter section, even before actually seeing the planet through an eyepiece.

Though my decision was ill informed, I soon learned that it had been a good one. It became clear at once that visual observations of the planet's clouds were dauntingly challenging since there was so much to be seen. With persistent effort, however, I gradually progressed to the point where I began to get some useful results. Over the next six years, that same small telescope faithfully helped me in drawing Venus, Jupiter, Saturn, sunspots, and lunar craters. It even provided an initiation into the basics of electronics when I first tried to build a motorized tracking unit for it.

My telescope had other unexpected benefits; it was small enough to fit easily in the family car, and it changed my summer vacations for good. After spending many summer days on the Mediterranean coast, it was time to forget about the beach and head for the Alps. There during the early eighties, I fell in love again with the pristine landscape and high-altitude meadows, the jagged rocky peaks, blissful silence, and crystal clear skies. Visions of nightlong darkness, without any artificial light, and reveling in the bright Milky Way suddenly became very real and as compelling and attractive as the imposing mountain wilderness around me.

Having a small telescope and being able to set it up quickly anywhere I chose opened up the new frontier of deep-sky observing. Much to my surprise, I also discovered that a telescope was a great way to socialize. In those remote mountain locations, my only company on mild summer nights were some local farmers and their families and a few vacationers. Still, all of them were eager to look through the telescope. In just a matter of minutes, one could get away from outdoor lights and switch on the Milky Way. Not much serious astronomy followed, but we had many laughs and good company to enjoy the sky above us. That's also where I started to appreciate that it is possible to establish a link between ordinary laypeople and the cosmos, often with unexpected results. The most typical and touching example to me was my grandfather, a retired farmer and wine reseller with little formal education. Looking at Jupiter or Saturn in awe, his most frequent exclamation was "Who knows why all these worlds are so round?!" Such a simple but profound and disarming question—whose explanation in simple terms, even after my studies in physics, would have required much preparation.

The amateur association William Herschel of Torino, whose weekly meeting I started to attend regularly, organized a yearly summer camp at 2,000 meters altitude, where each member would bring their own equipment for visual or photographic observing. I had so much fun there, eating good food and spending many nights with friends, making it impossible not to progress and become familiar with the sky, the constellations, and observational techniques—doing so with others added to the fun and motivation. Hunting faint nebulae, clusters, and galaxies was like doing one's own visual survey of the entire summer sky, and it became an absorbing activity each year for over a week of vacation.

A significant realization I made at that time was that the so-called theoretical limits of telescope performances outlined in textbooks were just that—theoretical. Even a small telescope, if used appropriately, could show diffuse objects so faint that I would not have believed seeing them was possible beforehand. Star hopping and averted vision (to exploit the peripheral sensitivity of the retina) became a standard for me. Eventually, I learned about visual narrow-band filters, and I started to take some pictures with cameras and lenses mounted piggyback on the telescope. I began getting good results with manual guiding, a taxing procedure that requires several minutes of staring at the same star in the eyepiece, while a firm hand gradually turns the knob of the telescope mount. Mostly I took wide-field photos of the summer Milky Way with color slide film, but I also started to use the legendary high-resolution, black and white Technical Pan 2415 film, after it was hypersensitized by a fellow member of our association. This prompted me to develop my films at home and experiment with different chemicals and photographic procedures.

Since our family lived in an apartment and not a darkroom, I had to buy dark fabric to cover the bathroom windows. I also discovered that I could transfer films to the development tank much faster by hiding everything, including myself, under my bedcovers. After a generous friend gave me an enlarger, photography became even more challenging and far more fun. There

FIGURE 11.2 The author observing the Sun with his 15-cm Newtonian, from 2,600 meters above sea level, in the late eighties. In the background is Mount Viso (3,841 m), which overlooks the large plain of the river Po across northern Italy and belongs to a UNESCO biosphere site along the border with France.

was much debate between astrophotographers who preferred light-toned images of nebulae, giving an effect that approximated what one saw visually in the telescope, and astrophotographers who overdeveloped the core of, say, the Andromeda galaxy to bring out subtle features like the dust lanes and faint spiral arms. I favored the trendier high-contrast approach and printed my images accordingly, though by today's standards, they would look crude at best.

Except for occasional mountain trips and vacations (figure 11.2), I was city bound. That, however, provided no serious impediment to doing what I enjoyed most, observing the brighter planets, where in contrast to deep-sky objects, there was always more than enough light available. I had now gained the experience needed to see and sketch many of Jupiter's cloud details. I bravely undertook *drift*, or *strip*, sketching, a technique long favored by observers of the Jupiter Section of the British Astronomical Association, in which one exploits the rapid rotation of Jupiter to add details as they cross the planet's central meridian (figure 11.3). This is a lengthy process. Continuously drawing cloud features for some hours as they slowly drifted across

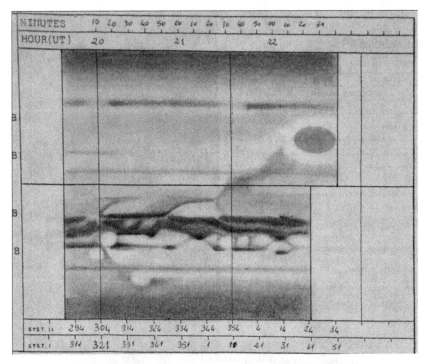

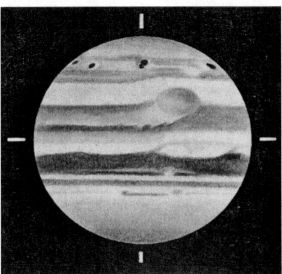

FIGURE 11.3 *Top*, a strip sketch of Jupiter, showing details visible in a 15-cm reflector as they passed across the central meridian of the planet over two and a half hours (a quarter of one Jupiter rotation). *Bottom*, the Shoemaker-Levy 9 impact scars as seen on July 21, 1994. Courtesy of Paolo Tanga.

the planet's disk required patience and perseverance, and it was fascinating to see this gradually unfold on paper. Moreover, by estimating the time when certain features cross the meridian, their position can be accurately computed. I was fascinated (and still am) that with practice, the eye and brain can develop the capability to measure speed and position of some variable phenomena with surprising precision, without the help of sophisticated technologies. I also embarked on other challenges, through visual observations of variable stars using the Argelander method (comparing apparent magnitude differences relative to nearby reference stars). My light curve of V781 Tauri was largely obtained under less-than-ideal conditions, on a freezing night during a particularly cold winter, while standing on a solid layer of ice that had formed on the terrace, where I was observing. While my friends were struggling with photoelectric photometry, I had more fun with my simpler approach, which I also applied in 1986 while observing mutual phenomena of Jupiter's satellites. By eclipsing and occulting each other, they rapidly fade away and reappear, as they pursue their endless pirouettes around the planet.

From such observations, rewarding for their own sake, valuable data regarding the physical and dynamical properties of the Jovian moons can be derived. I also got in touch (by the now forgotten traditional mail) with astronomer Jean-Eudes Arlot of the Bureau des Longitudes in Paris, who headed an observational campaign for such observations and sent me some informational material. Little did I anticipate then the life-changing events to come from my interest in those phenomena. But one thing was clear: observing the sky with a goal, or to measure or witness something new, really appealed to me much more than admiring or taking pictures. Documenting what I was observing made far more sense to me while spending hours under the stars. In spring 1986 my telescope grew larger and much improved in all aspects, with a new mount and a 15-cm mirror figured by a well-known Italian manufacturer, Romano Zen in Venice. Although not a big jump in size, this telescope was of such optical quality as to be serious business.

Writing and Studying

The mideighties were memorable in other respects. I was really fond of reading what other amateurs were doing, in Italy and around the world. Much of my curiosity was constantly fed by the monthly magazine *l'Astronomia*, edited by Corrado Lamberti, a high-school teacher with a PhD in physics, under the supervision of the celebrated Italian astrophysicist Margherita Hack. A couple of generations of amateur astronomers (and also some professionals) could vouch for how much that magazine influenced them. By treating astrophysics in all of its aspects (from history to literature, techniques to recent discoveries, debates, etc.), it inspired the brain and soul of numerous readers. Experts of international renown in all areas of astronomy were invited to write popular articles, thus resulting in a very high standard of quality.

Most of the journal's success was due to the kindness, expertise, and motivation of its director, Corrado. When I met him at an astronomy fair, I offered to write an article for him on amateur observations of Jupiter. That was the beginning of a long association with the magazine, as from time to time I was called on to write other articles, mostly on planetary observations. Those were my student years in physics at the University of Torino. They were intense, with new duties and friends, new dreams and challenges, but observing the sky and hiking in the Alps remained my preferred escape routes. While my studies were naturally trending toward astrophysics, over time my interests broadened, and I realized that other branches of physics were equally attractive to me.

For instance, those were the years when the study of chaotic systems and new descriptions of nature (through fractal geometry) started to be widely discussed. My fascination with this area led to my undertaking research on the transport of particles in turbulent fluids for my master's thesis. In the process I started to learn about the *coherent structures* (such as vortices) that sometimes appear in fluids. In the back of my mind, the link with the spots that I observed on Jupiter was always sharply present, with the Great Red Spot simply being the largest and most enduring of the vortex features forming and dissolving in the turbulent atmosphere of that giant planet.

New Friends, Mars, and More

My passion for observing the planets did not go unnoticed. Starting in 1986, some members of the national UAI established separate sections for all areas of planetary observations. While the sections for Jupiter and Saturn already had coordinators to collect contributions from amateurs across the country, there was really no such organization for Mars or Venus.

The initiative came from a bright and skilled amateur from Florence, Marco Falorni. Working on legal services at the Consiglio Nazionale delle

Ricerche (National Research Council of Italy), or CNR, he knew and often collaborated with professional astronomers at the Observatory of Arcetri, which overlooks the magnificent landscape of Florence and surrounding hills, very close to the last house where Galileo Galilei lived. As an expert visual observer of the planets and gifted with the best drawing skills, he was regularly given access to the observatory's 38-cm refractor. In addition to an attractive personality and a keen sense of humor, Marco also had a deep interest in ancient civilizations. Moreover, thanks to command of the full richness of the Italian language that only Tuscans seem to have, he readily shared his experience and views on the importance of planetary observations when done accurately and systematically, much as other famed planetary observers, G. V. Schiaparelli and E.-M. Antoniadi, had done.

I met Marco for the first time at a national meeting, and then on a hot summer day in Florence, where he invited a small group of persons designated to become the science committee of a newborn Planets Section of the UAI. I remember meeting in the cloister of San Lorenzo church, where old stones and vegetation created a microclimate well suited to our reflections.

This was the first opportunity for me to talk about planetary observations and to get in touch with a local group of astronomy friends who were experimenting with the most advanced techniques of photography and would pioneer amateur CCD imaging of the planets a few years later. Being part of such a group was not only rewarding for me, but also a confirmation that I was on the right path: establishing a rigorous approach for visual observations based on Marco's philosophy, as well as a way to convey to other amateurs the real flavor of scientific research. In this domain, approximation and incomplete data do not pay off. On the other hand, when observations are properly documented and compiled, they can form the basis of valuable research data. All this was still possible to do with small telescopes and pencil drawings, even at the beginnings of planetary exploration by space probes.

Spurred by this growing planetary fever, I decided to ask for access to a well-known 42-cm refractor at the Astronomical Observatory of Torino, to follow the "great" apparition of Mars in 1988, when Mars was both quite near Earth and favorably positioned for Northern Hemisphere observers. The director, who was also a professor in astrophysics at the faculty, introduced me to one of the few persons capable of handling this instrument, which at the time was used mainly for double-star measurements. Being a semiapochromatic triplet, it produces the cleanest, sharpest images of the planets that I have ever seen through any telescope.

Around the time of opposition, Marco had access to the Arcetri's refractor and was invited to join a small international team coordinated by Audouin Dollfus (another great visual observer, pioneer of polarimetry in astrometry and of high-altitude observations) to observe Mars with the Grande Lunette of Meudon. This was the same instrument that the great E.-M. Antoniadi had used in 1909, showing clearly the real nature of the "canals" of Mars. Our own observations were also satisfactory, and I was personally lucky to enjoy a few nights of nearly perfect seeing. Marco produced a precise map showing the south polar cap regression and collected a vast trove of observations from observers both in Italy and abroad. I had the opportunity to use the refractor of the Turino Observatory on several occasions later on. One of the most exciting in 1994 was following the expanding scars left in the atmosphere of Jupiter by the impact of the fragmented comet Shoemaker-Levy 9. Views of the planets, comets (Hale-Bopp!), and the Moon though that telescope remain among my fondest memories of the time.

While Marco took leadership of the Mars program of the national section, I was assigned Saturn. Despite giving the impression of a rather static globe, Saturn has subtle intensity variations in its atmospheric bands and zones that I found interesting to follow. I was also lucky enough to follow and study the emergence of one of the dramatic white spots, which seem to develop roughly once every thirty years, whenever Saturn's north pole is tilted maximally toward the Sun. The first one was Hall's Spot, noted in 1876 by Asaph Hall, soon to become famous for his discovery of the two tiny moons of Mars, with the 26-inch Clark refractor in Washington, D.C. Barnard's Spot appeared in 1903, and then came—perhaps the best known of all—Hay's Spot in 1933, so called because it was first recorded by W. T. Hay, the popular British stage and screen comedian, using a 6-inch refractor at his private observatory at Norbury, outside London. As with previous white spots, it rapidly lengthened toward the east over a period of weeks, and by the time it finally broke up, it had stretched halfway around the planet. Another white spot appeared on schedule in far northern latitudes in 1960. Between December 1989 and February 1990, Saturn's north pole was again tilted maximally toward the Sun, leading to expectations that another outbreak would occur—and so one did. In September 1990, an equatorial white spot was discovered by an amateur astronomer, Stuart Wilber, in Las Cruces, New

Mexico. It went from being a small intensely brilliant cloud, easily visible in a 2-inch refractor, to an oval cloud fifteen thousand kilometers long, elongated under the influence of the strong equatorial winds. (Note: since the north pole of Saturn was again tilted maximally toward the Sun in 2017, at the time of writing—February 2023—we would seem to be overdue.—W. S.)

In 1992, I began a regular collaboration with the journal *l'Astronomia*, through a monthly column on planetary observations that kept me busy for several years, until my real and "serious" job began to absorb more energies and time. In 1993, Marco, Corrado, and I managed to put together and publish a handbook for planetary observations that would remain a standard reference for amateur astronomers in our country for a long time. In addition to devoting many pages to visual methods, the handbook explained extensively how to properly observe the planets and record useful information, as well as the main reasons to study each planet, all elements still valid today.

Toward Formal Research (1992–1995)

Since my master's work was focused on the transport of solid particles in fluids, the next obvious step involved applying the mathematical simulations I had developed to my PhD research. Since I lived in the Piedmont area, I went to Torino for these studies—Torino, forever associated with the great name of Lagrange, whose university has boasted three Nobel laureates in medicine and physiology who trained under the Italian anatomist Giuseppe Levi: Salvador Luria, Renato Dulbecco, and Rita Levi-Montalcini. My own specialized research was facilitated by the ever more powerful computing facilities and expertise available in the Cosmo-Geophysics Institute in Torino, where I started my work for my PhD. I won't go into the details of the computing revolution I witnessed there (including the birth and rise of widespread computing resources), because that would deserve a full chapter of its own. It is worth mentioning, however, that this went in parallel with the ongoing expansion of space exploration and with my personal research trajectory.

I had bought my first cheap computer in 1984, secondhand from a high-school friend. It used audiotapes for permanent data storage, its display was a connection to a TV set, everything was black and white (there were no gray tones or colors), and the resolution was extremely low. Still, I was able to learn the basic principles of programming with it. During my master's and PhD work, I helped develop multiuser servers and had to install the Linux operating system from an endless set of floppy disks. I was also able to use Cray supercomputers, whose processors at the time had the power of a single modern laptop core. For personal computing, Apple I and II appeared, and then in a computer shop one day, I saw the first Macintosh, with its strange appendage (the mouse) and an interface based on clickable windows and buttons. I would have never imagined such a thing was possible. As a result of these innovations, my father's job as a book editor quickly became automated thanks to Macintosh capabilities.

In the early nineties, I used the first primitive browsers to access the novel and surprising World Wide Web, which quickly became one of the main sources for astronomical news and played a seminal role in disseminating predictions of comet Shoemaker-Levy 9's impact on Jupiter. When recalling the computers of Apollo 11 (less speed and memory than a four-operations pocket calculator) and the space shuttle (extremely reliable, but with less speed and memory than early generation smartphones), I can't help but see the link between space exploration and the progress in computers in general, all of which I experienced firsthand. After all, running numerical simulations to me was like creating a small fraction of a toy universe, whose richness and limitations were tightly bound to the power of the computers. This way I explored the trajectories of stratospheric balloons in the polar vortex. I also modeled the accumulation of dust in vortices of protoplanetary disks, which can trigger rapid planetary formation at discrete locations. The satisfaction of publishing my first scientific articles was great, especially illustrated with the almost artistic-looking plots that computers were capable of generating.

Clearly this new world was far removed from amateur astronomy, though the latter remained the chief occupation of my mind and hands whenever I was not working on my PhD. Specifically, some friends and I began to build a 40-cm Dobsonian telescope, which would become one of my traveling companions. I also participated with another group of people in the construction of a CCD camera from scrap. This home-built aluminum and electronically cooled camera head held a sensor that would be too small for even the cheapest camera today. The electronics were impressive though and entirely wire wrapped by one of my dearest friends, Saro Pomillo, using the expansion cards of a bulky "portable" computer. He also programmed the communication protocol between the camera and the computer. Although crude by today's standards, the camera turned out to be usable, but more important was the satisfaction of building it with friends and using it as an effective learning process. In 1993, I spent a few months with these friends at the Laboratoire de Météorologie Dynamique in Paris. I really enjoyed the opportunity to live as a real Parisian for a while, discovering the town but learning French the hard way, as I hardly had a very basic knowledge of it.

Anyway, while in Paris doing research on turbulent flows, I read in a scientific journal the predictions for a forthcoming season of mutual events among Saturn's satellites. It occurred to me that I could initiate a link between Italian amateur astronomers and the Bureau des Longitudes in Paris, not anticipating that they, too, would soon become close collaborators and friends. The following years were complex and eventful. I was nearing the end of my PhD and starting to think about what my next steps could be. Jean-Eudes Arlot of the Bureau des Longitudes invited me to a meeting in Bucharest, to plan an observing program of the mutual events of Saturn's moons. It became obvious to me that the division between amateur and professional activities was becoming more and more blurred. Indeed, I had now reached the point in my career where I was now meeting people I had read about, or with whom I had only corresponded by letter or email.

A pivotal moment for me occurred in summer 1995, when I read the book *Planets and Perception*, by William Sheehan (one of the editors of this book), which was a revelation to me, as it brought new perspectives, and not only from a historical point of view. Gradually, I realized how visual observations of the planets were in fact deep and meaningful experiments, ranging from the physical nature of the observed bodies to the mechanisms of human vision. Light from the planet crosses the turbulent layers of our atmosphere, then tests the quality of optics, and last triggers a complex process in the eye-brain system of an observer, with all the intricacies of a subjective interpretation. Sheehan's book clearly addresses this complexity, from a historical perspective and with concrete application to the "canals" of Mars. (I had never paid much attention to the fact that the astronomer who discovered the canals in 1877, G. V. Schiaparelli [figure 11.4] was born just ten kilometers from my hometown; he spent most of his career at the Brera Observatory in Milan.)

Years later, I realized how subjectivity and personal interpretation can affect science, well beyond the specific domain of visual observations. That same year, 1995, the death of Marco Falorni, after a short illness, sadly and

FIGURE 11.4 Giovanni Schiaparelli, discoverer of the "canals" of Mars, still scans the skies in this arresting sculpture in the Piazza di Indipendenza, in his hometown of Savigliano, only ten kilometers from where Paolo Tanga grew up. Courtesy of William Sheehan.

unexpectedly took our friend and great mentor away. I was chosen to lead the Planets Section after him, a particularly heavy task under the circumstances.

Closing the Loop (1996–2000)

The years of comet Hale-Bopp and immediately following were again times of changes and evolution. I was not fully convinced about a postdoctoral stay far from home, so I applied for a position as technician at the Astronomical Observatory of Torino, which today is owned and operated by the Istituto Nazionale di Astrofisica (National Institute for Astrophysics), or INAF. I did this almost on a whim and was surprised on hearing that I got the job, mainly because of the technical expertise I had acquired during the construction of the CCD camera.

Although I was formally assigned technical tasks, given my PhD and my scientific background, the director allowed me to join one of the science teams at the observatory and undertake scientific research. In a short time, this led me to a full-time job working in an area of my choosing: planetology. Transitioning from modeling simulations of turbulence to studying asteroids would not be easy for me. When I started my new job in summer 1996, I was somewhat overwhelmed by the weight of the challenge. I was also captivated, however, by the distinguished history of the institution, its many research efforts, and its impressive hilltop telescopes.

I earlier mentioned the observatory's 42-cm refractor. It was now just a short daily walk from my office. The other notable telescope was the 1-meter reflector in a special astrometric configuration. The old library, with its wooden shelves and large tables, was another highlight of the historical building. For the first time, my approach to astronomy had to evolve radically from a rather leisurely research pace in physics and outside interests. Astrophysical research now took on a totally different perspective: it was my job. Although I had not really sought or anticipated this transformation, here it was. I took it as a sign of destiny, demanding but clear. In the following years, I was driven full speed into a new realm, learning about the evolution of the asteroid belt, the collisional evolution of the solar system, the risk of impactors, and the many questions still unanswered.

I also got access to telescopes I had never used before and to powerful new research techniques. For example, I was able to use the fine guidance sensors of the Hubble Space Telescope in its interferometric mode to precisely measure the shape and size of asteroids. I also learned polarimetry by applying it to the dusty shells of comet Hale-Bopp. I traveled to Canary Islands to observe trans-Neptunian objects (TNOs) with the Telescopio Nazionale Galileo in the early phases of its operation and started to regularly attend international meetings. In Torino, I heard about the Gaia mission for the first time, something that I would encounter again in the future.

At the same time at the observatory, I took advantage of experiences gained during my nonprofessional years in preparation for observations during public events. I also managed to sign an agreement between the observatory and my amateur association for a more systematic use of the 42-cm refractor. This included full renovation of the tube and a refurbishment of the tracking mechanism. That marked a new start for the instrument, which remains today an outstanding facility for public events. At that time, I also enjoyed hosting week-long summer camps for high-school students, to develop some activities and learn about the work of astronomers. This was a great opportunity to interact with young people, whose eyes sparkled with that sense of curiosity about the stars I knew so well.

In 1997, Arlot and colleagues at the Bureau des Longitudes in Paris (by then renamed Institut de Mécanique Céleste et de Calcul des Éphémérides) invited me for an observing run of mutual phenomena of Jupiter's moons at the Observatoire de Haute Provence, advantageously located in one of the most attractive areas of Southern France, including lavender fields, postcard landscapes, and villages. Despite the intense summer heat, I enjoyed observing with a 0.8-meter reflector, again confirming the important role that telescopes of moderate size can play in planetary science. Most importantly, I met my future wife there, whose smile and eyes immediately became new bright stars for me.

Moving (2001)

I was fortunate to have forged growing links with France, for I began to find progress in my career difficult in Torino. A full-time research position there seemed elusive, so for the first time, I began to think seriously about a more promising location in which to continue my career. I already had contacts and collaborations with the Observatoire de la Côte d'Azur (OCA) in Nice. This facility included the old Nice Observatory, located on the summit of Mount Gros, featuring ravishing views overlooking the French Riviera, which in the late nineteenth century briefly enjoyed bragging rights as the residence of the world's largest refractor, a 77 cm by Henry and Gautier. The building was designed by Charles Garnier, and the main dome by Gustave Eiffel (figure 11.5). This awesome telescope was also—by several meters—longer than its rivals, the 76 cm at Pulkovo Observatory in the Russian Empire and the 68 cm at Vienna Observatory. Its long focal length was designed to optimize its performance on the planets during the heyday of the furor over the Martian canals discovered by Schiaparelli. Though some confirmations of these features were obtained by then director Henri Perrotin in the 1880s, the giant telescope seems to have generally disappointed expecta-

FIGURE 11.5 The dome of the Nice Observatory's 77-cm Henry and Gautier refractor, which in the latter nineteenth century was briefly the largest refractor in the world. Though the telescope is rarely used, the building, designed by Charles Garnier, with the main dome by Gustave Eiffel, remains an architectural jewel. Courtesy of William Sheehan.

tions, and its results were surpassed by those obtained with the 91-cm Lick refractor in California and the 83-cm "Grande Lunette" of Meudon.

As a scientific institution, the Nice Observatory no longer exists but was merged with the Centre de recherches en géodynamique et astrométrie (CERGA) in 1988. Then, in 2004, CERGA was dissolved, during a reorganization of its parent institution, the Côte d'Azur Observatory. When the opportunity came for a postdoctoral position there, I applied and was successful. This led me to move away at last, in February 2001, from the places where I had grown up.

FIGURE 11.6 The author doing a check of the 1.2-meter telescope at the Observatoire de Haute Provence, May 2021. This particular observing run was devoted to finding candidate new asteroids that had been observed by the Gaia mission just hours before. Though participating in making such observations is great fun, the author's job these days is mostly at the computer. Courtesy of Paolo Tanga.

Initially, my research and publications continued to focus on turbulence in early protoplanetary disks, specifically regarding the formation of high-density clusters through gravitational collapse within the gaseous protoplanetary nebula. Finally, in 2003, in the face of some strenuous competition, I was awarded a permanent position as astronomer at the observatory. I owe this success, of course, to the indispensable support of my new colleagues (figure 11.6).

In the early 2000s, the Gaia mission of the European Space Agency became a concrete undertaking, and Nice was one of the leading centers in its preparation, under the enthusiastic and expert leadership of François Mignard (figure 11.7). Gaia was conceived as a mission of fundamental astron-

FIGURE 11.7 François Mignard in 2010. Courtesy of William Sheehan.

omy. Over an operating period of five years, it was devoted to mapping the position, motion, and physical properties of stars with unprecedented accuracy. By continuously scanning the sky, Gaia would also collect observations of hundreds of thousands of asteroids. I soon became very interested in Gaia and began an amicable and close collaboration with François.

In 2005, I began working with the Centre national d'études spatiales (Center for Space Studies, the French space agency), or CNES, to plan the development of software for processing solar system data obtained by Gaia. A year later I led a team of about twenty people across Europe in designing and testing the software on simulated data. It was clear by then that a long, challenging, and engaging adventure had begun. Coordinating all this, participating in countless meetings, and conducting research all at the same time was so absorbing that I had to gradually give up some other activities, like writing articles and coordinating the Planet Section for amateur astronomers in Italy. One of my last roles was editing the national journal of the UAI from 2001 to 2003. After that, family life and the birth of another bright star, my daughter Chiara in 2008, were clearly deserving of most of my dedication outside work.

Though professionally, the early 2000s were mainly devoted to preparing for the Gaia mission, among the most remarkable events of these years involved the rare transits of Venus. Those of June 8, 2004, and June 5, 2012, were the first such events to occur since the nineteenth century, and they will not be repeated until 2117 and 2125. Needless to say, extensive observations were attempted around the world, and for the 2012 transit, I was fortunate to collaborate with solar astronomer Thomas Widemann of the Observatoire de Paris in developing the Venus Twilight Experiment, to probe the mesosphere of Venus by means of observations made with special coronagraphs. The latter had been custom built in Nice and were to be deployed at observing sites, including the Mees Solar Observatory in Haleakala, Hawai'i; Mobile Station in Hokkaido, Japan; Moondara Observatory at Mount Isa, in Queensland, Australia; Tien Shan Astronomical Observatory, Kazakhstan; Taiohae, Nuku Hiva, Marquesas Islands; Udaipur Observatory in India; and Lowell Observatory, Flagstaff, Arizona, in the United States. Each coronagraph was equipped with a different filter (blue 450 nm, visual 535 nm, red 607 nm, and infrared 760 nm); at Flagstaff I used a 535 nm to obtain CCD images, while William Sheehan used the same filter to obtain visual observations (to allow comparison with historic observations, such as those by the Russian academician M. V. Lomonosov in 1761; figure 11.8).[4] Though some of the stations were clouded out, the data collected allowed us to model refraction in the multilayered atmosphere of Venus, and the resulting vertical profile can be usefully compared to those from Venus-orbiting spacecraft as well as those derived from exoplanet observations.[5]

When Gaia was launched in December 2013, I was lucky enough to be invited to the event in French Guiana. Another momentous event was the release of Gaia solar system data on April 25, 2018: the first accomplishment after nearly fifteen years of study and work. Along the way, I have continued to publish about the evolution of asteroid shapes and the presence of primitive materials in some asteroid types, as well as enjoying a large network of collaborations. I have also tutored several PhD students and been responsi-

4. See, for example, J. M. Pasachoff and W. Sheehan, "Lomonosov, the Discovery of Venus's Atmosphere, and Eighteenth-Century Transits of Venus," *Journal for Astronomical History and Heritage* 15, no. 1 (2012): 3–14.

5. C. Pere, P. Tanga, T. Widemann, P. Bendjoya, A. Mahieux, V. Wilquet, and A. C. Vandaele, "Multilayer Modeling of the Aureole Photometry During the Venus Transit: Comparison Between SDO/HMI and Vex/SOIR Data," *Astronomy and Astrophysics* 595 (2016): A115.

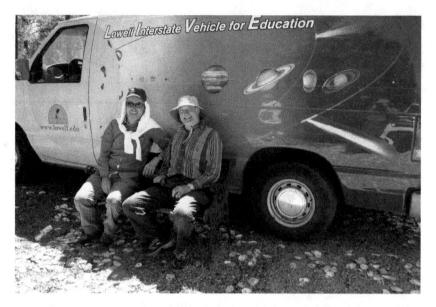

FIGURE 11.8 The author, *left*, and William Sheehan relax in front of a Lowell Observatory education-outreach vehicle at Lowell Observatory, Flagstaff, Arizona, after completing observations of the twilight arc of Venus as part of the Venus Twilight Experiment, June 5, 2012. Courtesy of Klaus Brasch.

ble for several postdoctoral fellows in Nice (some of whom got permanent positions afterward), all of which is ongoing.

Currently I am busy trying to build a network of collaborators to observe stellar occultations by asteroids more systematically, using a brand-new robotic telescope built by OCA for this purpose. With the accuracy brought by Gaia measurements of asteroids and stars, the occultation technique has now come of age and has great prospects for the future.

Final Thoughts

My account covers nearly half a century of curiosity about the sky, during which the first fuzzy images from the Apollo mission have been replaced by crystal clear close-ups of other worlds in our solar system. Over this time, my first attempt to compute asteroid positions by a pocket calculator have been replaced by large computing servers and online services running in "the cloud." The telescope in my backyard has grown more than three times in diameter (and much more in weight), with a plethora of accessories now readily accessible on the market. Today, space telescopes like Gaia provide unprecedented accuracy in our knowledge of the sky, and they will remain a cornerstone for decades to come. The sharpest telescopes start to probe the surroundings of stars and produce images of distant planets, while a new technique reveals merging black holes by detecting their gravitational waves. And more is expected from instruments of future generations.

It has been an amazing experience to participate in this grand adventure. But, at the same time, I continue to exploit small instruments that can bring the sky closer to everybody and even contribute to planetary science (among other sciences) in surprising ways. The telescopes used to measure the sunlight refracted in the Venus Twilight Experiment in 2012 were only 11-cm refractors. Many more 20- to 40-cm instruments regularly monitor distant solar system objects, revealing atmospheres and ring systems, and providing useful data to drive interplanetary space probes to them.[6]

That said, it is useful to add precision, to eliminate any possible ambiguity. Professional astronomy is different from the amateur approach. My rewards as a professional come today from having ideas to explain observed phenomena, from discussing them with colleagues, and from publishing in peer-reviewed journals. There is much effort, some stress, a lot of interactions and teamwork, strict deadlines, all taking place within an extremely complex academic environment. Nevertheless, I owe much to the heritage of my amateur astronomer side, which remains different and complementary to my job. Collecting photons in my retina that came from nearby planets or from remote galaxies was (and is) a priceless experience of the vastness of the cosmos, as well as a formidable starting point to question myself about our place in the universe and the *whys* and *hows* of our existence. It is also an endless source of cultural enrichment, linking present-day observers to the difficulties of the first ones, but also to the progress of techniques and physical knowledge about the universe through the ages, up to our contemporary times. Astronomy is one of the rare disciplines in which amateurs can make concrete contributions to professional science, and they do that so well by

---

6. The New Horizons (NASA) mission was capable of obtaining useful data from the flyby of the TNO Arrokoth thanks to occultations observed in advance. An occultation campaign on the asteroids that are trojans of Jupiter drives the preparation of the Lucy mission. Near-Earth asteroids, too, have been studied first by occultations for other space missions (DESTINY+ of the Japanese space agency JAXA, and DART/Hera of NASA/European Space Agency).

finding new ways to push the instrumental limits and reach better results. I cannot forget my experience, suggesting that astronomy is really for everybody and a great way for people to find out firsthand what science is. I also cannot forget people who, over time, have made my interest in astronomy meaningful. Some of those I mentioned, such as my grandfather, my uncle (first owner of the famous handbook), Corrado, and Marco, are no longer with us. All of them were essential, as are the many friends and colleagues I have today, to shaping my own way of looking at the sky.

In the end, only one element remains to be added. My perception of the sky as a pristine component of nature, at the same level for me as mountain wilderness, came in parallel with my consciousness of the duty we have to preserve the beauty of our planet, to protect it from overexploitation and pollution. Having lived for a long time in a big town, I experience this feeling directly.

Today, space exploration seems to increasingly turn into exploitation and business, while humanity faces unprecedented challenges for preserving the environment and ensuring our own survival. In the coming decades, increasing knowledge of the other planets and exoplanets, including how they evolved, can help us t understand the danger we run into, as well as our place in the universe. But in our everyday activity as astronomers (professionals and amateurs alike), we can help spread awareness of the beauty of the sky, as fragile as wilderness on Earth. It is probably our duty and a small contribution to keep alive the sense of wonder and the protection that nature deserves.

My Way to Space Studies

ALEXANDER T. BASILEVSKY

F ar from being direct, my way to space studies was full of digressions and unplanned zigzags. After graduating from high school in early summer 1954, a time when young people often select what their future should be, my schoolmate and friend Volodya Shibanov and I dreamed of becoming sailors. Since we were both nearsighted, however, that career path was not an option. Instead, we decided to become geologists and applied to take entrance exams in the Department of Geology of Voronezh State University. (Voronezh is a center of Voronezh province, about five hundred kilometers south of Moscow, and probably best remembered for the attack in June 1942 that represented the first phase of the German army's 1942 campaign in the Soviet Union. During the war, the city was almost completely destroyed and had to be rebuilt.) Besides the romance of becoming geologists, there was another attraction: student-geologists were to be issued impressive looking military-type uniforms, complete with golden epaulets.

After successfully passing entrance exams in six subjects, Volodya and I were admitted to the university. Studies were to begin on September 1, 1954. The day before, there was an organizational meeting for new students, during which, along with information on class schedule and other things, we were informed that the USSR Ministry of Higher Education had canceled the uniform for student-geologists. One of the main attractions of becoming student-geologists was thus taken away. Naturally, as newcomers, we were quite disappointed, but it was too late to do anything about it. We managed to borrow uniforms from older students, however, and to pose in photos (figure 12.1).

FIGURE 12.1 *Left*, the building the Department of Geology shared with some other departments in the early fifties at Voronezh State University; *right*, a photo of the author in the canceled student-geologist uniform. Courtesy of Alexander T. Basilevsky.

Volodya and I found the lectures and seminars in geology immensely stimulating, and at the beginning of summer after our first year, we began to receive training in the practical, hands-on side of things by doing some field geology among the sedimentary rocks near Voronezh, southwest of town. In fact, these sedimentary rocks, which belong to the Voronezh Massif, are rather famous. As is now widely accepted, the massif is a tectonic anteclize, which formed mostly during the Hercynian orogeny (a mountain-building event about three hundred million years ago during the Late Paleozoic continental collision between Euramerica and Gondwana to form the supercontinent of Pangaea). The bedrocks there are Precambrian and sometimes rise to the surface of the Earth; the sedimentary rocks are of Riphean and Vendian age, and the southern portion consists of Devonian, Permian, and Triassic cover, along with some Mesozoic and Cenozoic rocks. All in all, a great area in which to begin our training—though in the Soviet Union at the time (early fifties), almost all the leading geologists and geophysicists were opposed to the idea of continental drift.[1]

---

1. V. E. Khain and A. G. Ryabukhin, "Russian Geology and the Plate Tectonics Revolution," in *The Earth Inside and Out: Some Major Contributions to Geology in the Twentieth Century*, vol. 192, ed. by D. R. Oldroyd, 185–98 (London: Special Publications, 2002).

Field practice continued the following summer in Crimea, with visits to the Black Sea shore and the folded Crimean Mountains, which drop steeply to the Black Sea toward the southeast. After our third year, we began serious fieldwork as members of geological teams in different regions of the Soviet Union. I was in a small party studying ore deposits in Central Kazakhstan when I heard my first summons to space. On October 4, 1957, in the company of a few colleagues, I was celebrating my twentieth birthday. While drinking something a little stronger than tea, we were listening to the radio and heard that the Soviet Union had launched Sputnik 1. It may seem surprising, but I was absolutely indifferent to this news. We were in a geologically very interesting place, with beautiful mineralogy of wolframite-molybdenite ores. That's what I was focused on. As for Sputnik, my reaction was: Who cares about this?

After graduating from university, I joined a geological mapping expedition in the central region of the European part of the USSR. Although the work was pretty interesting and proved useful in my subsequent career, I later learned, much to my surprise, that some key issues, like the ages of the mapped geological units, had not yet been definitely established. Since I was drawn to more quantitative methods of research, I decided to become a PhD student in geochemistry. After intensive self-education in such mandatory topics as geochemistry, the English language, the history of the USSR Communist Party, and philosophy, I successfully passed my preliminary exams and in 1963 became a PhD candidate in the Geochemistry Division of the Department of Geology at Moscow State University. My adviser was Alexander Vinogradov, an academician (i.e., a member of the prestigious Academy of Sciences of the Soviet Union). He was head of geochemistry, director of the V. I. Vernadsky Institute of Geochemistry and Analytical Chemistry,, and would become something of a legend—twice receiving the title Hero of Socialist Labor (1949, 1975), and having a mountain on the near side of the Moon and a large crater on Mars named in his honor. The work I was assigned under his supervision was devoted to experimental studies of melting crystallization in the olivine-chromite system.

As the experimental techniques available at Moscow State University were not adequate for my work, Vinogradov invited me to the Vernadsky Institute, one of the major research institutes of the Academy of Sciences, which Vinogradov had led since its founding in 1947. After World War II and during the early Cold War period, work at the institute was primarily focused on atomic power production, which was highest priority at the time. This included research in radiochemistry, separating trans-uranium elements, and prospecting for the raw materials, such as uranium, needed for atomic power production and, implicitly, for use in nuclear weapons. Since that time, the institute had greatly broadened the scope of its research. On my arrival, I became friends with Andrey Ivanov, who, under the direction of Cyril Pavlovich Florensky, was involved in the study of cosmic spherules in various sediments, including those in the region of the so-called Tunguska catastrophe. This 1908 event in Siberia was widely regarded as the result of a small cometary impact. Florensky was the son of one of the twentieth century's most intriguing Russian intellectuals, Pavel Alexandrovich Florensky. As a Russian Orthodox theologian, priest, philosopher, mathematician, physicist, electrical engineer, inventor, polymath, and neo-martyr, he has been referred to as "Russia's unknown da Vinci."[2]

In 1966 my time as a PhD student expired, and I returned to the geological mapping expedition and became involved in some further geochemical studies. Since my doctoral dissertation was not finished yet, I worked on it evenings and during my days off.

A sudden and dramatic change in my life occurred in the early summer of 1968, when I got a call from Cyril Pavlovich Florensky, inviting me to join his Division of the Moon and Planets at the Institute of Space Research of the Academy of Sciences. Vinogradov, who was vice president of the Academy of Sciences at the time, had transferred him to the Institute of Space Research to help select landing sites for a Russian crewed expedition to the Moon, which was being planned under project N1-L3. For this important project, Florensky was allocated some fifteen open positions, for which he invited young people with very different expertise, including volcanology, geochemistry, mineralogy, petrology, soil science, and geodesy-cartography. His thinking was that since this was a new area of research, we could work as a multidisciplinary team and teach each other in whichever area we had expertise. Our task was to select landing sites and to provide the spacecraft engineers with information about craters and boulders on the lunar surface to help them design a lander with the necessary capabilities to arrive safely.

---

2. A. Pyman, *Pavel Florensky: A Quiet Genius—The Tragic and Extraordinary Life of Russia's Unknown da Vinci* (New York: Continuum International, 2010).

For this work we were provided a set of high-resolution photographs of the lunar surface taken by the U.S. Lunar Orbiter missions in 1966–67. Since the Soviet Union had no such high-resolution imagery of the lunar surface, Soviet reconnaissance services were asked to obtain them from the United States, and they did. I do not know whether the photos provided for us were bought, gifted, or stolen, but we used them for our work. Using tracing paper and clear plastic over the images, we outlined and counted craters, rocks, and boulders and classified them into different morphological categories. Usually, work of this type is very boring, but in our case, it was not, because every Friday some engineers from OKB-1 would visit us, take new sets of information on the spatial distribution of craters and rocks, and use this quantitative data for more detailed understanding of the lunar surface. While we were working on the Lunar Orbiter images, our director, Cyrill Florensky, and his deputy, Alexander Gurshtein, with money from OKB-1, enrolled several astronomical observatories in the USSR in support of the N1-L3 project. Their data were used to estimate lunar surface roughness and gain better positional information (coordinates) of features. This intensive work lasted for about a year and resulted in our report to OKB-1 with information on spatial and surface characteristics of four landing sites in the equatorial zone of the Moon.

Using data gained through this work, we published a major paper describing typical characteristics of the surface of lunar maria.[3] Meanwhile, after cancellation of the U.S. Apollo program at the end of 1972, the Soviets also saw no further reasons to pursue the N-1/L-3 program (which had been expensive and unsuccessful so far). In May 1974, Leonid Brezhnev, acting on recommendations of the Central Committee secretary, Dmitri Ustinov, replaced Vasily Mishin with longtime rival Valentin Glushko at OKB-1, and though at the time, two N-1 launch vehicles were nearly ready by the end of the year, Glushko suspended the program. Two years later, the nearly finished N-1s and components were destroyed, and the N1-L3 was decisively terminated. By then the Soviet lunar program had abandoned the dream of crewed landings and instead redirected to a series of innovative and successful robotic probes.

Led by Georgy Babakin at the Lavochkin Association, ten spacecraft, starting with Luna 15 (which crashed onto the Moon while the Apollo astro-

---

3. C. P. Florensky et al., "On the Problem of Characteristics of the Surface of Lunar Maria," in *Modern Concepts on the Moon* (Moscow: Nauka Press, 1972), 21–45 (in Russian).

nauts were on the surface) and ending with Luna 24 in August 1976, reached the Moon. Among these, Luna 15, 16, 18, 20, 23, and 24 were designed as sample-return missions, Luna 17 and 21 brought rovers to the surface, and Lunokhod 1 and 2 were intended to investigate lunar surface characteristics via several scientific instruments. Luna 19 and 22 were orbiters whose primary task was to improve our knowledge of the lunar gravitational field. Our division in the Institute of Space Research was responsible for selecting and characterizing landing sites for the sample-return missions and the Lunokhods. We had already gained invaluable experience through similar work for the N1-L3 project, but because of the ballistic-engineering constraints of the Luna landers, not all of it was applicable. The landing areas were outside the zone for which we had the high-resolution Lunar Orbiter photos, so we had to essentially depend on the astronomical observations.

Luna 15, the first of these missions, was intended to return samples of lunar materials back to Earth. The Lavochkin Association provided us with coordinates of the area within which we should select a landing site with a relatively smooth surface. The area was in the southern part of Mare Crisium, and using the information provided to us by the Kharkov Astronomical Observatory, we selected the landing site. The Luna 15 mission began on July 13, 1969, and three days later, NASA launched the Apollo 11 mission, intended to land in Mare Tranquillitatis, making it a true space race. On July 17, Luna 15 entered a circumlunar orbit and on July 21, after two orbital corrections, was transferred to the landing trajectory. Unfortunately, 267 seconds after the start of this operation, the radio connection with Luna 15 was suddenly lost, ending the mission. Meanwhile on July 20, Apollo 11 landed successfully and started its way back to Earth the following day, landing in the Pacific Ocean on July 24.

Following the Luna 15 failure, the flight engineers blamed those of us who had selected the landing site, claiming that we had overlooked a mountain into which the spacecraft crashed. Consequently, we had to show our photographic and cartographic evidence to academician Mstislav Keldysh, president of the Soviet Academy of Sciences, and convince him that no mountains were present within the prescribed landing ellipse. After confirming that this was the case, he put an end to the unjust attacks on us and allowed us to resume our everyday work. The Luna 16 mission began the next year, and on September 12, 1970, it successfully landed in Mare Fecunditatis and returned some 101 grams of lunar soil. The landing site for this mission was selected by our group in the Lavochkin Association. A year later, Luna 18 was launched, with the goal of returning soil samples from the highlands between Mare Fecunditatis and Mare Crisium. Unfortunately, it too crashed during landing, but this time we were not blamed for the catastrophe. In February 1972, however, Luna 20 successfully landed in the same highland region and brought back 55 grams of lunar materials. Then in October 1974, an attempt was made to obtain samples from Mare Crisium with Luna 23. It had improved drilling equipment aimed at obtaining core samples to a depth of about 2 meters. Sadly, the lander overturned after and could not complete its mission. And finally, in August 1976, the Luna 24 mission to the Moon successfully landed in Mare Crisium and brought samples in the form of 160-cm long cores weighing about 170 grams.

Two months after the flight of Luna 16, Luna 17 was launched and successfully landed on November 17, 1970, in the northwestern part of Mare Imbrium, where it deployed the robotic rover Lunokhod 1. Luna 21 was launched in January 1973 and landed in the crater Le Monnier, which has a mare-like floor and thus is a bay of Mare Serenitatis. It deployed Lunokhod 2. I was part of the team that selected the landing sites for both missions. During the Lunokhod missions, a group of geologists and cartographers from our team, including me, worked in the command center near Simferopol, Crimea, from which the Lunokhod crew remotely controlled the operations on the lunar surface. We used panoramas and images taken by the rover navigation TV cameras in combination with data obtained by other scientific instruments of the mission (figure 12.2).

FIGURE 12.2 *Left*, part of the TV panorama taken by Lunokhod 2 in crater Lemonier. *Right*, the author and Boris Nepoklonov, head of the scientific group of the Lunokhod missions. Courtesy of Alexander T. Basilevsky.

Lunokhod 1 had covered a distance of about 10 kilometers on the Moon's surface by September 1971, when the temperature inside the instrumental module became too low during the lunar night. The rover had several scientific instruments onboard, including panoramic TV cameras, an X-ray fluorescence spectrometer RIFMA to measure the elemental composition of the surface, and a PrOP instrument to measure the load-bearing capacity and shear strength of the upper 5–10 cm of lunar soil. In addition, it was equipped with an X-ray telescope to measure radiation coming from the sky, and a French-made laser retroreflector to measure variations in the Earth–Moon distance, indicative on the internal structure of the Moon. We did not have high-resolution images of the Lunokhod 1 study and therefore never knew what we might encounter several meters ahead of it.

Lunokhod 2 carried the same set of scientific instruments but was also equipped with a magnetometer to measure local magnetic fields and an upward-pointing astrophotometer to measure the brightness of the lunar sky and thereby reveal the presence of dust suspended above ground. At the start of this mission, we lacked sufficiently detailed photos of the study area. Thankfully, noted planetary geologist Harold Masursky of the U.S. Geological Survey visited Moscow for a scientific conference and brought us high-resolution images of the crater Le Monnier taken by the Apollo 15 orbital module. These provided us with a better understanding of the local lunar surface and allowed us to plan our route to reach interesting objects to study.

Unfortunately, our working time with Lunokhod 2 was unexpectedly short. On lunar day 4, Lunokhod 2, while driving with an open solar battery, suddenly entered a fresh crater about five meters in diameter, and while attempting to exit, its solar panel touched the crater's inner wall and became soiled. This was immediately apparent since the onboard battery's electric current decreased abruptly. At the end of the lunar day, it became necessary to cap the solar battery and avoid supercooling the rover's radiator during the frigid lunar night. By doing this, however, the radiator became covered with lunar soil, which, it turned out, is a very effective thermal insulator. The following lunar morning, the radiator was covered by soil and consequently radiated heat very poorly, causing the spacecraft to overheat. That unfortunately ended the mission prematurely, but during its short working time, Lunokhod 2 traveled about thirty-nine kilometers, almost four times more than Lunokhod 1.

In the early seventies our division of the Space Research Institute, working with data obtained by the Lunokhods, continued to study the Lunar Orbiter images and examine the published results of the Apollo missions. I was able to combine my knowledge of surface processes, obtained from analysis of Lunar Orbiter and Lunokhod imagery, with measurements of the surface exposure age of rocks on the rims of small craters in Apollo landing sites and publish an article on the dependence of absolute age of craters on their size and morphological maturity. At that time as well, the Space Research Institute was assigned a new director, academician Roald Sagdeev. He declared that "real" science is represented by mathematics and physics, so he moved the "nonscientific" divisions of this institute to other institutes. As a result, we moved to the Vernadsky Institute of Geochemistry and Analytical Chemistry, where we were renamed the Laboratory of Comparative Planetology.

In the late seventies, our laboratory became involved in evaluating data from the Soviet missions to Mars and Venus. Of two spacecraft launched for Mars in July 1973, Mars 4 was a failure, but Mars 5 successfully entered orbit around the planet in February 1974. Although only partly successful, due to a gas leak from the instrument module, the orbiter was able to gather data for nine days. It returned sixty images of the Martian surface, which added some new information to that obtained by the U.S. Mariner 9 mission of 1971–72, and obtained data with the gamma-spectrometer, allowing us to measure thorium and uranium content over vast regions, representing two major types of Martian surface: (1) volcanic plains and shield volcanoes, and (2) cratered highlands.

On the other hand, the Venera missions (Venera means Venus in Russian) were among the USSR's greatest space achievements. In 1972, Venera 8 landed on the planet's surface and measured the content of U, Th, and K in its top layer. Veneras 9 and 10, consisting of both orbiters and landers, reached Venus in 1975. The landers successfully reached the surface and returned close-up images of the landing sites, as well as measuring the U, Th, and K content of the surface layers by gamma-spectrometry. In 1978 Veneras 11 and 12, equipped with TV cameras to obtain close-up images of the surface, failed to do so because the camera covers did not separate after landing. Veneras 13 and 14 successfully landed on Venus in March 1982 and obtained images of the surface close to the landing sites, as well as measurements of the rock-forming elements by X-ray fluorescence spectrometry. Several members of our group, including me, studied images taken by Veneras 9,

10, 13, and 14, and found that all four locations exhibited finely layered sedimentary rocks and loose soil. Although the origin of those layered rocks was not known at the time, after later analysis of NASA's Magellan radar images, we suggested that these might represent atmospheric sediments originating from the fine-grained fraction of ejecta of large impact craters forming so-called radar-dark parabolas.[4]

In the eighties, after an extensive review of planetary science literature, combined with the Laboratory of Comparative Planetology's accumulated information on the geological characteristics of the various planets and satellites, lab members, led by Cyril Florensky, published two major books: *Essays of Comparative Planetology* and *Impact Craters on the Moon and Planets*. In 1982 Cyril Florensky passed away, and I started to take over his duties as head of our laboratory. This did not, however, take much of my time.

The 1983–84 Venera 15 and Venera 16 mission orbiters undertook lateral radar surveys of the northern quarter of Venusian surface and provided images of 1–2 kilometer resolution and surface elevation information with spatial resolution around 5 kilometers. Analysis of these images by our laboratory and a couple other Soviet science institutions led to identification of the major geological formations and structures of this planet (figure 12.3).

Beginning in the late seventies, academician Valery Barsukov, then director of the Vernadsky Institute, started to organize visits of Soviet planetary scientists to the United States to attend Lunar and Planetary Science Conferences in Houston. We typically arrived at Kennedy Airport in New York, where we would be met by Toby Owen, professor at Stony Brook University, and stay in Stony Brook for a couple of days while adjusting to the local time. From 1985 till 2009, Brown University and the Vernadsky Institute organized so-called Vernadsky-Brown micro-symposia: two per year—one in Providence, then in Houston, in March just before the Lunar and Planetary Science Conference, and another in Moscow in October—for a total of fifty. During our stays in the United States, we visited various geologically interesting places and some scientific institutes, for example, Meteor Crater and the nearby USGS Astrogeology Science Center in Flagstaff, Arizona.

---

4. A. T. Basilevsky, J. W. Head, and A. M. Abdrakhimov, "Impact Crater Air Fall Deposits on the Surface of Venus: Areal Distribution, Estimated Thickness, Recognition in Surface Panoramas, and Implications for Provenance of Sampled Surface Materials," *Journal of Geophysical Research* 109 (2004): E12003, 5002.

FIGURE 12.3 *From left to right*, James Head, Alexander Basilevsky, Valery Barsukov, and Lionel Wilson discussing images of Venus's surface taken by the Venera 15 and 16 missions. Courtesy of Alexander T. Basilevsky.

In 1989 two other Soviet scientists and I were sent to the U.S. Jet Propulsion Laboratory as guest investigators in the Voyager 2 mission. At that time Voyager 2 flew by the giant planet Neptune, and in collaboration with American colleagues, I studied the high-resolution images of Triton, the largest (D = 2,700 km) satellite of this planet. For a couple of months during that visit, I was able to identify linear features on Triton's surface reminiscent of strike-slip faults, indicative of strong horizontal tectonic stress in the crust of this body.

In 1989 the American spacecraft Magellan successfully orbited Venus and scanned it by radar from 1990 to 1994. It covered almost the entire surface of the planet to lateral resolution of about 100–200 meters, including topographic detail with horizontal resolution of 5 kilometers. Since geologists and cartographers from our laboratory already had experience working with the Venera 15 and 16 radar images of Venus, they also participated in analyzing the Magellan data. At that time and later, we had a very productive collaboration with James Head, of Brown University, both by visiting Providence, Rhode Island, and by working at the Vernadsky Institute. My personal contribution to that collaboration was to produce a model of

FIGURE 12.4 *Left to right,* Anne Cote (wife of James Head), James Head, the author, and the German planetary scientist Harald Hiesinger, in the vicinity of Providence, Rhode Island, April 2012. Courtesy of Alexander T. Basilevsky.

the global stratigraphy of Venus.[5] This model was used and applied in several regional geological maps as well as in a global geological map of Venus. In addition, my Vernadsky Institute colleagues collaborated with James Head and his team on various aspects of planetary geology, a collaboration up to the present time (figure 12.4).

At the end of the eighties, Peter Janle from the Institute of Geophysics at the University of Kiel, Germany, invited me to visit his institute, which led to our collaboration in studies of the geology and geophysics of Venus based on analysis of data from the Pioneer Venus Orbiter and Venera 15 and 16.

Since the nineties our laboratory has also collaborated with Gerhard Neukum from DLR Institute of Planetary Exploration, Berlin, Germany. This collaboration became more intensive when Neukum became principal investigator for the European mission Mars Express, which launched in 2003, carrying, among other instruments to study Mars, a high-resolution stereo camera. After he became head of the Institute of Geological Studies at the Free University of Berlin, I frequently visited him, resulting in several joint publications on various aspects of Martian geology. During the late nineties and early 2000s, I gave a series of short courses on planetary geology at Oulu University in Finland and collaborated with Jouko Raitala, leader of their planetary geology group, and his PhD students. We analyzed the Magellan radar side-looking images of Venus, which resulted in several joint articles.

In 1999 I was invited to visit the Max-Planck-Institut für Aeronomie, in Katlenburg-Lindau, Germany, for a joint photogeological analysis of TV images taken by the Mars Pathfinder lander and the small rover Sojourner. The landing site was in Ares Vallis, a large outflow channel. We found some specific features in the local landforms indicative of water flow, which traversed

5. A. T. Basilevsky and J. W. Head, "Regional and Global Stratigraphy of Venus: A Preliminary Assessment and Implications for the Geologic History of Venus," *Planetary and Space Science* 43, no. 12 (1995): 1523–53.

the Ares Vallis. In 2004 the institute partly changed its research directions and was renamed the Max Planck Institute for Solar System Research; then in 2014 it moved to the city of Göttingen. My collaboration there has continued until recently. We studied not only Mars, but also the Moon, Venus, and comet nuclei, including comet 67P Churyumov-Gerasimenko. Our Venus studies were based on analysis of data obtained with the Venus monitoring camera onboard the European Space Agency (ESA) Venus Express. I was involved in the analysis of thermal emission data of the Venusian surface in the so-called atmosphere transparency window, where we found several relatively small areas with temporary increases in thermal emission. These are located in one of the young rift zones on the planet, and the most probable explanation for this phenomenon is volcanic eruptions.

Another interesting finding made jointly with colleagues at the Max Planck Institute for Solar System Research was identification and study of the so-called pinnacles on the nucleus of comet 67P Churyumov-Gerasimenko. They are topographic promontories first observed on the nucleus of comet Wild 2. Their characteristics suggest that they are erosional remnants created by a loss of the surrounding material. Comet 67P was a subject of studies by the ESA mission Rosetta. Analyzing images of the comet's nucleus, I noticed the promontories, which reminded me of the pinnacles on the Wild 2 nucleus, and began a study involving my Vernadsky and German colleagues. We found that the features of the 67P pinnacles agree with the suggestion that they are erosion remnants formed as a result of a slowdown of erosion in places where nucleus material is more resistant to subliminal erosion than the material around it. In this case, our measurements showed that the maximum heights of the pinnacles (100–200 m) gives a lower boundary for the amount of surface material lost, and their diameters (typically tens of meters) are a measure of the size of the erosion-resistant parts.

In 2016 I was invited to Macau University of Science and Technology, in Macau, China, to give a two-week course of lectures on planetary geology, and I did the same in 2017, 2018, and 2019. These invitations were made by Wing-Huen Ip, professor at the National Central University of Taiwan, who also taught part time at Macau University. We got acquainted in the early 2000s when he was working at the Max Planck Institute for Solar System Research while I was visiting this institution. Professor Ip asked me to be the adviser of his PhD student Yuan Li, who was studying survival times of rock boulders on the lunar surface. My advisory work and collaboration with

Yuan Li resulted in a couple of joint publications as well as her successful PhD defense. Now Dr. Yuan Li is on the staff of Suzhou Vocational University in China, and we continue to collaboratively research the geology of the south polar region of the Moon.

During my visits to Macau, I became acquainted with young scientist Feng Zhang. Using very high-resolution LROC NAC images, he was identifying and mapping the spatial distribution of small volcanic domes in lunar maria, each rimmed by a moat, or a ring-moat dome structure (RMDS). A few hundred meters in diameter and only a few meters high, RMDSs look morphologically fresh and are evident in many regions of maria with surfaces approximately three billion years old. We know that mare basalts are covered by layers of regolith four to five meters thick. If RMDSS were as old as the surrounding basalts and only a few meters high, they should not be visible and certainly not morphologically as sharply defined as they are. Consequently, they must be very young, indicating that we must reevaluate the history of lunar volcanism. I was a member of a small international group organized by Feng Zhang, and we have published several joint articles.

What I have written here shows that I have had the pleasure of collaborating with many foreign scientists and institutions, and that these collaborations have taken place over a significant portion of my scientific career. I am thankful to the Vernadsky Institute for its philosophy of encouraging such international cooperation.

13

A Spark for Astronomy

LEO AERTS

I was born in 1951 and grew up in rural Flemish Belgium, some fifty kilometers from the capital, Brussels, and near the port city of Antwerp. At a very early age, I was drawn to all aspects of nature, biology, geology, physics, but especially astronomy, and enchanted by the beauty of the night sky, simply because it was there (figure 13.1). This was still possible a decade after World War II, when light pollution didn't really exist in Western Europe. That is when my curiosity about astronomy was sparked, and it has remained with me ever since. Today, I can look back over six decades of progressive understanding of both theoretical and observational astronomy, but what have not changed in the least are my ongoing wonder about and fascination with the nature of the cosmos and our place in it.

In the midfifties, although the dire economic situation in Belgium began to improve thanks to the U.S. Marshall Plan, there was still no information or educational opportunity related to astronomy in the rural setting where I lived. Consequently, my curiosity and youthful hope to learn more about the stars went unfulfilled. Thankfully, all that changed in the sixties, when local newspapers began publishing daily articles on science and sometimes even information about forthcoming astronomical events. When Mariner 4 revealed craters on Mars in 1965, I was truly excited, not so much about the news itself, but by the fact that the promise of space travel had clearly begun to be fulfilled. The planets were within reach now, and even those remote stars seemed less distant to me. From that point on, astronomy, especially the observational side, became an integral part of my life.

FIGURE 13.1 Despite growing up in an impoverished, poorly educated small rural village in northern Belgium, the author was struck by the first sparks for astronomy even as a four-year-old boy (fifth from left, back row). Courtesy of Leo Aerts.

In the early sixties, with the improving European economy, life became easier, even in rural areas. After a stint in the Belgian army, I was able to focus once again on the things I really cared about. Before knowing anything about astronomy and telescopes, I eagerly looked for any news in local newspapers about the American and Soviet space probes. I also discovered books by Patrick Moore, which became available in Belgium at the time. Unlike many older amateurs, I did not fully appreciate the long-standing debate about the origins of lunar craters that Moore and others felt so strongly about, namely volcanism as opposed to impact cratering. Although aware of those debates, I was much more interested in obtaining more information about other aspects of astronomy.

Even more exciting to me really were all those magazines with thrilling stories I had never heard of before—science fiction. Since these cheap paper publications were affordable, I eagerly entered the fascinating realm of "exploring new worlds and going where no one had gone before," all well before Captain Kirk and Mr. Spock appeared on television. In fact, almost nobody in Belgium had TV at that time, which only became widely available in the late sixties. Instead of amusing myself passively by sitting in front of the TV, I expressed my fascination with space travel, astronauts, and imaginary worlds through my own drawings (figure 13.2).

Things really began in earnest for me in January 1973. I bought my first quality telescope, a 4-inch Unitron refractor, and joined the Vereniging Voor Sterrenkunde, a Belgian association of mostly amateur astronomers. After the last Apollo landing, in December 1972, and cancellation of further flights, the Moon became less interesting to most people, including many amateur astronomers, who no longer perceived it as enticing so much as an irritating celestial beacon that interfered with other astronomical observations. For me, however, it had the opposite effect—the drama of the Moon landings motivated me to learn more about the only heavenly body accessible to detailed examination from Earth. The stars struck me as mere points of light, the planets as tiny disks, and comets mostly dim, short-lived phantoms in the sky. The Moon seemed far more inviting, as a world that one could almost touch, with innumerable and varied surface features to explore telescopically in sharp relief, shifting through endless cycles of lights and shadows. I did

FIGURE 13.2 Some of the author's drawings illustrating his early space age memories. Courtesy of Leo Aerts.

not realize then that I was embarking on a voyage of discovery that would last for a lifetime.

Instead of thinking that the Apollo missions had delivered the coup de grace to what could be learned from Earth about our natural satellite, I was determined to find out all I could about it with the instruments at my disposal, my eyes and telescope, and eventually through photography. As so many others before me had done, I began by examining and sketching the most interesting lunar features my telescope revealed, learning their names, and watching them change in appearance under varying angles of illumination (figure 13.3). In this way I first entered an enigmatic realm of unearthly landscapes, with a seeming infinity of pits and craters, meandering rills, smooth expanses of lava and imposing mountains, whose nature and origins, so long contested, was only now becoming better understood by geologists.

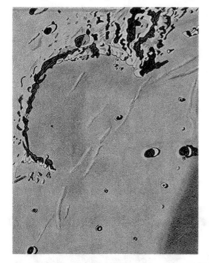

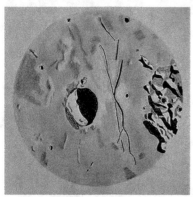

FIGURE 13.3 Examples of early sketches of lunar features that attracted the author's attention, including *top*, the beautiful Sinus Iridium (Bay of Rainbows), and *bottom*, intricate details of the Triesnecker Rimae region. Courtesy of Leo Aerts.

I first photographed the Moon with my 4-inch telescope in 1973, but after a decade in which this fine instrument satisfied me, I began, as many other amateurs have, to yearn for something larger, especially after seeing enticing telescope ads in books and magazines. Thankfully, I was now gainfully employed and had the means to purchase several different telescopes over time, including 6- and 8-inch f/15 refractors, large-aperture Newtonian reflectors, and eventually Schmidt-Cassegrain and Maksutov type instruments as well as several rich-field telescopes (figure 13.4). People not familiar with amateur astronomers will probably find all this a bit obsessive! I fully agree, but as obsessions go in my view, this is a relatively harmless one. Merely owning these instru-

FIGURE 13.4 Two of the author's favorite telescopes used for lunar and planetary observing and imaging efforts. Courtesy of Leo Aerts.

ments was never an end in itself, as each was put to good use in broadening my perspective of our remarkable universe and also in sharing this with others.

I have often been asked why I did not try to become a professional astronomer, since I was so interested in all aspects of space from early childhood. In the eighties I often visited the Royal Observatory in Uccle, Belgium, and relished reading many of the books housed in its beautifully appointed library, where I especially enjoyed delving into the history of astronomy. While there, I was always welcomed and never felt out of place, but more like a student eager to learn as much as he could about the universe for the pure joy of it. In short, I was a consummate amateur astronomer and never saw this as anything more.

Professionally I have had several diverse careers. When I was child, most of my peers typically left school at the age of twelve or fourteen years to work in factories. I was fortunate to be able to attend school until I turned eighteen, when I also began to work in a factory. That was really the only way to earn a living at the time in that still poorly educated part of my country. Subsequently, I worked for the Belgian Federal Police for several years, for the Department of Social Security, and as a quality control agent for acquisitions for the Belgian army. Last, I was a purchasing agent for security equipment in several financial institutions. Along the way, I have been happily married for fifty years to my wife, Simonne, and have one son, Tom, neither of whom share my interest in astronomy but "tolerate" my longtime passion for it.

As noted, my first youthful attempts to photograph the Moon proved quite challenging. Since my small refractor came with an altazimuth rather than a more expensive equatorial mount, it could not track the Moon but only take brief snapshots with my old Exacta VX500 camera and the grainy films available at the time. Still, I was overjoyed with those modest results, as the lunar maria and many terminator features were clearly evident. This proved a true eureka moment for me as a youngster and has influenced my astronomical trajectory to this day.

A revolution took place at the turn of the new millennium, as the digital era was born, and along with it, totally new technologies we could use and apply more effectively to astronomical imaging. Almost overnight, we did away with films and chemical processing, in favor of powerful new imaging and processing modes that completely overshadowed what had gone before. At last amateurs could do what seemed impossible before and capture images of the Sun, Moon, and planets far better than what was visible by eye and to the full potential of their optical equipment. With even more advances in digital technology since then, the work of amateur solar system imagers today approaches a quality obtainable before only through satellites and space probes. Thanks to the advent of webcams, for example, I was now also able to wander beyond the Moon and capture some truly satisfying high-resolution images of the Sun and planets and even the major moons of Jupiter and Saturn.[1]

Perhaps nothing illustrates this proverbial quantum leap in technological advancement more than my first attempt in 2003 to image Saturn with a webcam and a Celestron-14 telescope (figure 13.5). Well known for the subtlety and range of its pastel colors, very delicate cloud features, and of course its remarkable ring system, Saturn has always been not just a favorite but also a challenging planet to observe, sketch, and photograph for both classic and modern students of the solar system. Like most other amateurs after years of fuzzy film photos, I fully embraced this challenging new technology of astronomical imaging in ways not possible before. Today, with even better cameras and processing software, Mars too can be studied in remarkable detail with amateur instruments, especially during favorable oppositions (figure 13.6).

---

1. See, for instance, K. Brasch and L. Aerts, "Observations of Jupiter's Moons," *Astronomy* 48, no. 11 (2020): 46–49.

FIGURE 13.5 *Left*, the author's first webcam image of Saturn and Titan, taken in 2003. *Right*, Mars during the favorable 2020 opposition. Both were taken with a Celestron-14 telescope. Courtesy of Leo Aerts.

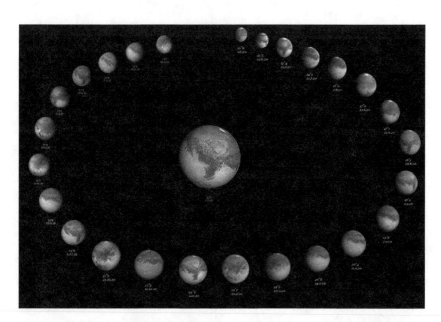

FIGURE 13.6 A montage of Mars images, June 1 to December 18, 2020, covering the entire period in which the planet's apparent diameter was greater than ten secs of arc. The opposition date was October 13, when the apparent diameter was 22.3 secs of arc. Incidentally, the circumstances were almost identical to those of the October 20, 1894, opposition of Mars, observed by Percival Lowell at Flagstaff. These webcam images were taken with a Baader IR filter and a Celestron-14 telescope. Courtesy of Leo Aerts.

Like so many lunar observers before me, I have been fascinated by such classic features as the prominent craters Tycho, Clavius, and Copernicus, the "monarch of the Moon," as well as the intriguing geology of the Hyginus rill and Straight Wall. I have been most awestruck, however, by the towering mountains seen in projection against the darkness of space along the southern limb of the Moon. Because of its fleeting visibility from Earth, this region has long haunted students of the Moon, and in addition to its majestic features, it is uniquely evocative in that one has the impression of actually orbiting the Moon in a spacecraft and looking out at its terrain through a porthole window. Consequently, whenever conditions are favorable, I have focused my attention on lunar limb regions.

I was, of course, far from the only observer to be so captivated, and perhaps I can be allowed a brief segue into selenographic history—with the assistance of selenographic historian William Sheehan.[2] As is well known, the Moon always keeps one side turned toward the Earth and the other turned away. This is not quite true; because its orbit is elliptical, inclined to the plane of the Earth's orbit, and disturbed in various complicated ways by the Sun, it rocks slightly back and forth and up and down, so that in all 59 percent of its surface can be seen at one time or other from the Earth.

Because of these librations, lunar limb regions are revealed only during certain favorable presentations and then appear in highly foreshortened form, so that topographic knowledge about these areas was long desperately limited. (The Soviet Luna 3 mission of October 1959 was historic in providing humans a first glimpse of the hitherto forever hidden farside regions, though as the images were very poor by later standards, the exact nature of most of the details could not be determined, with the exception of a few cases, such as the lava-flooded crater named Tsiolkovskiy.)

Of all the foreshortened regions of the libratory zones, none was more tantalizing to earthbound observers than that around the South Pole. Its awe-inspiring landscape was described by the inimitable British amateur astronomer Richard Baum as a realm of "craggy irregularity and elemental splendour," with "majestic peaks, some scalloped, others tooth-shaped, all rising against the pitch-black background to dazzle the eye—one can almost imagine them

---

2. William Sheehan, with Thomas A. Dobbins, of the magisterial history of selenography: *Epic Moon: A History of Lunar Exploration in the Age of the Telescope* (Richmond, Va.: Willmann-Bell, 2001).

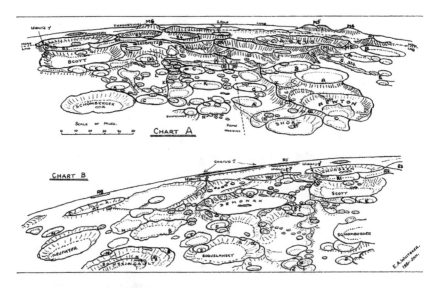

FIGURE 13.7 The towering mountains of the Moon's south polar region, which present only tantalizing views from Earth because they are highly foreshortened by perspective. These early maps were drawn by E. A. Whitaker in 1954 and were long the best available. Courtesy of E. A. Whitaker, Leo Aerts Collection.

to be mountains of Eternal Light, as in the imaginations of older observers."[3] No one who ever caught a glimpse of these regions can ever forget them. Several amateurs have spent significant parts of their careers carefully observing these regions and attempting to map them—an effort that retained its piquancy owing to the mischance that the south polar region was the only significant area of the Moon not included in coverage by the Lunar 4 Orbiter in 1967 (it lay in shadow at the time). Among these observers were Harold Hill, an inspector of mines by vocation and a highly accomplished draftsman, who embarked in 1951 on what would turn into an almost forty-year obsession with the south polar region; Ewen Whitaker (figure 13.7), later a key member of G. P. Kuiper's team at the University of Arizona, but in early days a moonstruck member of the Lunar Section of the British Astronomical Association, who studied the Moon in his spare time while employed in spectroscopy at the Royal Observatory, Greenwich, and produced an influential

---

3. Richard Baum, "Harold Hill and the South Polar Region of the Moon," *Journal of the British Astronomical Association* 120, no. 2 (2010), 86–97, quotation on 89.

chart of the south polar region in 1954; and John Westfall, a geographer by profession, who referred to the region as "Luna Incognita," and who would be the only one to complete and publish a map.[4]

The region has now been mapped in great detail by the Clementine and Lunar Reconnaissance Orbiter spacecraft, but it has not lost its allure to amateur observers seeing its magnificent jumbled terrain come into view, and it retains a romance not unlike that of polar explorers on Earth. It is a place of cherished dreams, long pursued and never completely fulfilled, of observers who without ever leaving the eyepieces of their telescopes found satisfaction wandering hither and thither among the vast scenery of another world.

Though I didn't personally contribute significantly to the mapping of the south polar region, I nevertheless have been a regular wanderer among this scenery and found, for at least for brief periods, that I have been able to forget the petty concerns of everyday existence while being transported into a transcendent realm, where (as the poet Keats said) "love and fame to nothingness do sink" (figure 13.8).

Today, of course, the Moon's polar regions have attracted a great deal of interest with the discovery by several orbiting spacecraft of water ice in their cold, permanently shadowed craters. Possibly delivered there over extended geological time periods by ice-bearing comets and asteroids, or by not yet fully established chemical mechanisms, significantly large deposits of water would be invaluable for long-term human habitation efforts on our nearest neighbor in space. It would also go a long way toward bringing my youthful fantasies of establishing human colonies on the Moon from mere science fiction to reality.

Einstein said at an official celebration of the sixtieth birthday of Max Planck something that I think captures the exaltation of all of us, professional and amateur alike, who have devoted themselves to spending their nights studying the distant worlds in the sky:

I believe with Schopenhauer that one of the strongest motives that leads men to art and science is escape from everyday life with its painful crudity and hopeless dreariness, from the fetters of one's own ever-shifting desires. A finely tempered nature longs to escape from

4. J. E. Westfall, "Mapping Luna Incognita," *Journal of the Association of Lunar and Planetary Observers* 34, no. 4 (1990): 149–59.

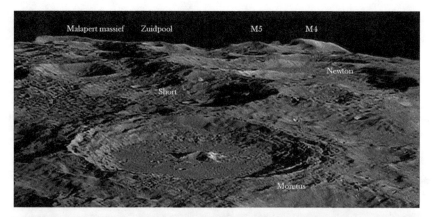

FIGURE 13.8 Some of the author's south polar region images from 2015. *Top*, the prominent crater Moretus in the foreground and the massive peaks M5 and M4 of the Leibnitz Mountains. *Bottom*, the region including the crater Shackleton that lies in permanent shadow and is now known to contain water ice. Courtesy of Leo Aerts.

personal life into the world of objective perception and thought; this desire may be compared with the townsman's irresistible longing to escape from his noisy, cramped surroundings into the high mountains, where the eye ranges freely through the still, pure air and fondly traces out the restful contours apparently built for eternity.[5]

As astronomers realize better than anyone, those high mountains, where the eye fondly traces out the restful contours apparently built for eternity, might as well be on the Moon as any place else.

5. Quoted in Banesh Hoffmann with collaboration of Helen Dukas, *Albert Einstein: Creator and Rebel* (New York: Plume, 1973), 221.

Some Concluding Thoughts

WILLIAM SHEEHAN

I n the introduction, I quoted Carl Sagan referring to knowledge of the solar system: "Clearly the best time to be alive is when you start out wondering and end up knowing. There is only one generation in the whole history of mankind in that position. Us."

Among the questions we have begun to answer is the nature of the "distant and indistinct discs," to borrow another Sagan phrase for the planets visible in the night sky, which were so long subject to little more than tracking of their motions. Even in the fifties, when many of the authors whose reflections are contained within this volume were coming of age, and the first rockets began to set out from Konstantin Tsiolkovsky's "cradle of humanity," we knew surprisingly little about the Moon and planets, and often what we thought we knew turned out to be wrong. There is nothing humiliating about that, by the way. It is the way of science. Space exploration has created a great example of how science uses observations to correct old ideas and move humanity forward. Since the fifties—when one author, like Patrick Moore, could cover in reasonably complete detail the whole waterfront of what was known of the solar system—the river of knowledge that was then a comparative trickle has become a tsunami. Thanks to our spacecraft emissaries and the instruments they have carried, we have now visited not only the Moon and planets but also asteroids, comets, and moons. They are no longer distant and indistinct disks, but known places, and their study has given rise (as our own world has) to vastly expanded scientific disciplines, including astronomy, geology, physics, chemistry, and biology, as well as a host of multidisciplinary ones. The amount of information is now so overwhelming that it is a cliché to say that no person can possibly comprehend the big picture, though it can be argued that the big picture involves backing off and getting the first-order details clear rather than ending up in the weeds, learning all the fourth-significant figures.

And what is the big picture? The canvas of the solar system (not even including the Sun, which is a subject entirely unto itself) is colossal, breathtaking in scope, and not easy to take in. The main features are now deeply researched, however—the way the planets formed from the nebula that enveloped the Sun, the bombardment from leftover debris that wreaked havoc across the early solar system, in which particularly large impacts took place, including one that bequeathed to us our faithful and useful companion, the Moon; the chemical differentiation of different worlds, in the terrestrial planets producing globes containing a nickel-iron core, an olivine-rich mantle, and a rocky crust, and in the giant planets' enormous spheres of hydrogen and helium, whose gases are so compressed in the interiors they assume liquid and metallic forms. We have studied varied geological processes, including cosmic impacts that formed the ubiquitous craters on many of these worlds. Likewise we have modeled how primordial gases belched out by volcanoes affected planetary climates and formed primitive atmospheres, worked out how processes of gentle heating, exposure to UV irradiation and intermittent drying can convert simple organic (carbon-based) chemicals into RNA, proteins, and other core components of cells. We have stood in awe of the gigantic swirling atmospheres of Jupiter and Saturn, wondered at the enormous tilt of Uranus on its axis, studied the intricate gravitationally induced patterns exhibited in ring systems (especially the unparalleled set possessed by Saturn), and added Jupiter's moon Europa and Saturn's Enceladus to possible worlds on which life might have formed. Indeed, understanding the origin of life finally seems within our reach, while semiquantitative estimates of the preponderance of life—and even intelligent civilizations—in the galaxy, though still highly speculative, are at least not entirely so.[1] Even perhaps the last frontier of science—understanding the nature of consciousness itself—is no longer hopelessly out of reach but has begun to

1. See, for instance, Tom Westby and Christopher J. Conselice, "The Astrobiological Copernican Weak and Strong Limits for Extraterrestrial Intelligent Life," *Astrophysical Journal* 896, no. 1 (2020): 58–75.

be accessible to the methods of science, though in that case, not because of anything we learned through the space program.[2]

Much of this has occurred during the last sixty or seventy years, and so in the lifetimes of many still living (though now inevitably passing from the scene). Most of the authors of the essays contained herein came to intellectual awareness and embarked on their scientific careers during the exciting years—the fifties, sixties, and seventies—when the Moon and planets first began to be explored by spacecraft, and planetary science, as its own discipline, was born. All the authors have possessed, to a rather unusual degree, the trait of epistemic curiosity—engaging experience, logic, and reason to systematically probe the unknown. Probably it is to some extent inborn. Yet, it can be encouraged—or discouraged. It can thrive, or perish by benign neglect—or downright suppression. Despite all the faults of our time, epistemic curiosity has been (at least until recently) valued and encouraged, especially in science. Journalist Ian Leslie reminds us of how atypical this has been, historically:

> Our oldest stories about curiosity are warnings: Adam and Eve and the apple of knowledge, Icarus and the sun, Pandora's box. Early Christian theologians railed against curiosity: Saint Augustine claimed that "God fashioned hell for the inquisitive."
>
> A society that values order above all else will seek to suppress curiosity. But a society that believes in progress, innovation, and creativity will cultivate it, recognizing that the inquiring minds of its people constitute its most valuable asset. . . . Curious learners go deep, and they go wide. . . . They are the ones most likely to make creative connections between different fields, of the kind that lead to new ideas and the ones best suited to working in multidisciplinary teams.[3]

In all of human history, the periods in which epistemic curiosity has been encouraged and thrived have been rare and short-lived (and for good reason have been regarded from the perspective of later times as golden ages). One such period occurred in ancient Greece between about the sixth cen-

---

2. For a lucid introduction, see Anil Seth, *Being You: A New Science of Consciousness* (New York, Dutton, 2021).

3. Ian Leslie, *Curious: The Desire to Know and Why Your Future Depends on It* (New York: Basic Books, 2014), xiv and xvi.

tury, when Ionian philosophers like Thales, Anaximander, and Pythagoras appeared, and the second century BCE, when Hipparchus on the island of Rhodes brought observational astronomy to a state of perfection it would not approach for another seventeen centuries. Another was in the Italian city-state of Florence during the Renaissance, when after the fall of Constantinople to the Arabs, ancient Greek texts began to be transported by Byzantine scholars to Italy, which had the effect of introducing diversity into the human mind by offering ideals of human thought and action different from those insisted on by the medieval church and would produce a Leonardo who in the 1480s produced a doodle in his notebook: "Tell me . . . tell me whether . . . tell me how things are."[4] That impulse of Leonardo—wanting to find out how things are—continued during the Renaissance, the Reformation, and the Enlightenment, when "received wisdoms began to be interrogated, and . . . European societies started to see that their future lay with the curious and encouraged probing questions rather than stamping on them."[5] That Enlightenment impulse has encouraged "the biggest explosion of new ideas and scientific advances in history," though it should be noted that not all societies involved have been democratic or open; Nazi Germany contributed more than any other nation to the development of rockets, while it was almost a fluke that the democratic United States beat the authoritarian USSR to the Moon.

It is beyond the scope of this epilogue to consider the implications of that, though it is worth noting that Sagan, in his last, and perhaps bravest, book, *The Demon-Haunted World*, published just before he died in 1996, expressed the following concerns:

> Science is more than a body of knowledge; it is a way of thinking. I have a foreboding of an America in my children's or grandchildren's time—when the United States is a service and information economy; when nearly all the key manufacturing industries have slipped away to other countries; when awesome technological powers are in the hands of a very few, and no one representing the public interest can even grasp the issues; when the people have lost the ability to set their own agendas or knowledgeably question those in authority; when, clutch-

---

4. Quoted in Leslie, *Curious*, xix–xx.
5. Leslie, *Curious*, xv.

ing our crystals and nervously consulting our horoscopes, our critical faculties in decline, unable to distinguish between what feels good and what's true, we slide, almost without noticing, back into superstition and darkness.[6]

When the contributors to the present volume were asked what had changed about their point of view in the years since they began their participation in the great adventure of exploring the solar system for the first time and establishing the multidisciplinary field (or fields) of planetary science, the most consistent themes were that though they were grateful to have lived through and contributed to all that, looking at the current situation, they all tended to worry about the future prospects of the scientific endeavor and even the long-term continuation of human life on the planet.

Brasch wrote, "After barely surviving World War II and growing up first in Rome and then Canada, it gradually dawned on me that I would become a scientist. Although first attending Catholic schools, I was never convinced that religion provided the answers I was seeking. . . . Instead, science provided me the broad perspective on the natural world that had first attracted me as a small child, while also, to paraphrase Tsiolkovsky, 'doing something useful . . . and propelling mankind forward, if only by a fraction.'" Although Brasch feels he did his part through his decades of scientific (biomedical) research, teaching and advising thousands of students during his active years as an academic, he now notes more ominously: "I am growing alarmed at our state of public education, the rise of superstition, mythology, and seemingly willful ignorance or denial of science in many sectors of society. And this at a time in history when we face truly existential issues like climate change, nuclear war, and overpopulation."

Cruikshank wrote, gratefully: "My devotion to astronomy began some years before the first artificial Earth satellite was launched, at a time when humans on the Moon and the exploration of the solar system were only dreams. Now, twelve people have walked on the Moon, more than two hundred astronauts have orbited the Earth in the International Space Station, and several asteroids and comets have been studied with robotic spacecraft. . . . I'm optimistic about the continued exploration of the universe, both

---

6. Carl Sagan, *The Demon-Haunted World: Science as a Candle in the Dark* (New York: Ballantine, 1996), 25.

locally and beyond. A permanent colony on the Moon and humans on Mars in my lifetime seem to be reasonable expectations, and I can hardly wait!"

He continues: "But if I were to indulge a darker view, I would lament the course that an alarming fraction of the world relentlessly has and continues to take, with wars, famine, and all forms of human suffering, the vast majority of which is avoidable and results from continuing ignorance, indifference, and utter folly. . . . As long as avarice, hubris, and egotism loom large in the human psyche, I am pessimistic about man's future."

Hartmann, looking back fondly on the kinds of questions that had fascinated him as a fourteen-year-old—the nature of the Moon and planets, whether they might be inhabited—felt that by the time he had reached midcareer, "I saw answers to my fourteen-year-old self's first-order questions emerging, sometimes in papers I had written or co-authored. But by the time I was in my fifties, sixties, and seventies, I was seeing that much of the culture around me did not accept the answers we scientists were finding. I began to be more interested in the ultimate mystery, humans themselves. . . . A constant 35–40 percent of American citizens accept a 'creationist' view (from the 1400s–1600s) that the Earth was created only about 6,000 years ago, that biological evolution never happened, and that consistent dating by various radiometric techniques, in various countries, of various rocks from Earth, Moon, Mars, and asteroids, all give false answers. My main concern these days is how to respond to our current situation."

As scientists nearing the end of our careers (and we suspect that younger scientists are similar), we are more comfortable describing the exhilaration of research and discovery than playing the Cassandra role of warning as society careens seemingly heedlessly toward potentially cataclysmic and civilization-ending results. Also, as scientists, we are used to having a long view, and to taking in stride the roughly 13.8 billion years since the Big Bang, the 4.5 billion years since the Sun and Earth-Moon system formed, and the 300,000 years since *Homo sapiens* appeared. One thing about science that sets it apart from other fields of human endeavor is that it is always unfinished, a work in progress. Its close ally includes not certainty (illusively offered by dogmatic religions) but doubt. Science is not even primarily a body of knowledge but, as Sagan said, "a way of thinking." He continues:

The scientific way of thinking is at once imaginative and disciplined. This is central to its success. Science invites us to let the facts in, even when they don't conform to our preconceptions. It counsels us to carry alternative hypotheses in our heads and see which best fit the facts. It urges on us a delicate balance between no-holds-barred openness to new ideas, however heretical, and the most rigorous skeptical scrutiny of everything—new ideas and established wisdom. This kind of thinking is also an essential tool for a democracy in an age of change.

One of the reasons for its success is that science has built-in, error-correcting machinery at its very heart. Some may consider this an overbroad characterization, but to me every time we exercise self-criticism, every time we test our ideas against the outside world, we are doing science. When we are self-indulgent and uncritical, when we confuse hopes and facts, we slide into pseudoscience and superstition.[7]

No matter how bleak things may seem at the moment, it is this ever-renewing resilient character of science that also provides some hope. Ideologies, rigid dogmas, authoritarian institutions, violent methods, which Isaac Asimov called "the last refuge of the incompetent," inevitably, over time, come into conflict with the way the world works and are prone to end, like the Third Reich in 1945, in smoldering ruins. Science by contrast is capable of self-renewal; it has a self-correcting mode built in that tests itself against the outside world and makes adjustments accordingly. As such, at least until and unless humans really do go too far and make the Earth and life on it part of the smoldering ruins of their failed ideologies, rigid dogmas, authoritarian institutions, and violent methods, science will always have the last word—because it never has a last word.

As scientists, we can only hope, with Isaac Asimov: "Humanity has the stars in its future, and that figure is too important to be lost under the burden of juvenile folly and ignorant superstition. The saddest aspect of life right now is that science gathers knowledge faster than society gathers wisdom."[8] Perhaps, in the end—as we as a society of humans struggle under the weight of more and more information, much of it spurious and unreliable "alternative facts" and amounting to little more than noise—we can affirm that wisdom does not lie in the fourth-significant figures but in getting the big

---

7. Sagan, *Demon-Haunted World*.

8. Isaac Asimov, *Isaac Asimov's Book of Science and Nature Quotations* (New York: Grove Press, 1990).

picture right, and that involves collectively using our knowledge and humanity to ensure survival on our planet. The big picture includes the realization that we are all stuck (eight billion of us at present) on the same planet, and that beautiful and interesting as the other worlds we have discovered are, they offer no immediate alternatives for us to escape to.

No; in the end we have only this planet to live on, and only science to guide us, which, for all its flaws, is better than the alternatives.

CONTRIBUTORS

Leo Aerts came of age in the late fifties, at the dawning of the space age, when he first became enthralled with the possibility of space travel and developed an all-consuming interest in the phenomena of the night sky. Among the activities in which he participated were variable star observations and the photography of the Sun, Moon, planets, comets, and objects in the deep sky. Although his professional career in the private sector had no connection with science, he has continued to be active in amateur astronomy, having served as director of the planetary and cometary section of the Flemish Astronomical Society Vereniging Voor Sterrenkunde (VVS), co-authoring a deep-sky atlas, and publishing many monthly columns for Dutch and Belgian astronomy magazines. He has been honored with the Galileo medal of the VVS for his many contributions to amateur astronomy.

Alexander T. Basilevsky had no interest in astronomy in high school or university where he completed his master's and PhD courses. This only developed when he began studying lunar morphology as a member of the team at the Space Research Institute of the Soviet Academy of Sciences charged with selecting and characterizing landing sites for the planned (but never realized) Soviet crewed expedition to the Moon. He found this work very interesting, and went on to work on the surface morphology/geology of the Moon, Venus, Mars, Phobos, Callisto, Ganymede, Triton and Mercury collaborating with scientists of the United States, Germany, Finland, and China. These fundamental science studies involved participation in the space mission teams and selection of landing sites for the Soviet, then Russian, space missions: Lunokhod-1, 2, Luna 15, 16, 17, 18, 20, 21, 23, 24, Venera 15/16. Basilevsky has published more than 250 articles in peer-reviewed journals and three books (in co-authorship). He has given semester-long courses on planetary geology at Moscow State University, as well as many short courses in different universities both Russia and elsewhere. Since 1975 he has worked at the Vernadsky Institute of Geochemistry and Analytical Chemistry of Soviet/Russian Academy of Sciences, returning in the last several years to again work almost exclusively on the morphology of lunar surface.

Klaus Brasch was fascinated by the natural world from a very young age, especially astronomy (as a member of the Royal Astronomical Society of Canada) and biology. Professionally, he opted for biology and proceeded to earn MS and PhD degrees in cell and molecular biology at Carleton University in Ottawa, whereupon he embarked on an academic career in biomedical and cancer research, though still retaining an avocational interest in astronomy. A highlight of the latter was joining with colleagues in developing and teaching advanced courses in astrobiology. He served as dean of science and director of campus research at California State University, San Bernardino, and was awarded Research Advocate of the Year 2004 by the U.S. Small Business Administration for facilitating the commercialization of university-derived technologies. After retiring to Flagstaff, Arizona, he has been active with the public program at Lowell Observatory, in astrophotography, and as a frequent contributor to *Sky & Telescope*, *Astronomy*, and the *Journal of the Royal Astronomical Society of Canada*.

Clark R. Chapman was interested in astronomy and journalism from an early age, and immediately after graduating from high school, he began measuring lunar craters at the University of Arizona. His choice of planetary science as a career was inspired when, as a Harvard undergraduate, he took a course from Carl Sagan. He studied Jupiter's atmosphere and asteroids at the Massachusetts Institute of Technology and carried out research at the Planetary Science Institute in Tucson from 1972 until 1996, when he moved to the Space Studies Department of the Southwest Research Institute in Colorado. He wrote a well-reviewed book, *The Inner Planets*, republished with the addition of Voyager spacecraft results as *Planets of Rock and Ice* (Scribner's), and chaired the Division for Planetary Sciences of the American Astronomical Society in the early eighties. He has served on the science teams of the Galileo, Near Earth Asteroid Rendezvous–Shoemaker, and MESSENGER Mercury missions. He was the first editor of *Journal of Geophysical Research—Planets*, co-authored (with David Morrison) the book *Cosmic Catastrophes* (Plenum/Springer), and received the Carl Sagan Medal for Excellence in Public Communication in Planetary Science from the American Astronomical Society. He has specialized in studying asteroids and the holes they make in planetary surfaces (impact craters), co-created the concept of asteroid rubble piles, and was an early advocate of protecting humanity from a catastrophic but very unlikely asteroid impact.

Dale P. Cruikshank is an astronomer and a planetary scientist. His research specialties are spectroscopy and radiometry of planets and small bodies throughout the solar system. He uses spectroscopic observations made with ground-based and space-based telescopes, as well as interplanetary spacecraft, to identify and study the ices, minerals, and organic materials that compose the surfaces of planets and small bodies. Together with several colleagues, Cruikshank has discovered many kinds of ice on several small planetary bodies and, with the Cassini spacecraft, hydrocarbons on three of Saturn's satellites. Cruikshank finished his graduate studies with a PhD in 1968 at the University of Arizona as a student of G. P. Kuiper. After a year in the USSR as a National Academy of Sciences exchange scientist, he moved to the University of Hawai'i in 1970, as a tenured astronomer at the Institute for Astronomy. Cruikshank joined NASA in 1988 and retired in 2021. He has been a member of the science teams for NASA's Spitzer space telescope and its Voyager, Cassini, and New Horizons missions. Cruikshank has served as an officer of the International Astronomical Union and the Division for Planetary Sciences of the American Astronomical Society, and he was elected a Fellow of the California Academy of Sciences, the American Geophysical Union, the American Astronomical Society, and the American Association for the Advancement of Science. He received two NASA medals for exceptional scientific achievement and the NASA medal for exceptional service. In 2006 he was awarded the G. P. Kuiper Prize of the Division for Planetary Sciences of the American Astronomical Society.

William K. Hartmann is known internationally as a planetary scientist, writer, and painter. In 1974–75, he originated (with Planetary Science Institute colleague Donald R. Davis) the now generally accepted theory of the origin of the Moon (involving a giant impact on newly forming Earth). He's known also for developing a system to estimate the ages of planetary surfaces by counting the numbers of impact craters per square kilometer. He served on the imaging teams of three different Mars orbiters, including one in Europe. With colleague Dale P. Cruikshank, he made many observations of asteroids and comets at Mauna Kea Observatory in Hawai'i, revealing that outer solar system asteroids and even ice-rich comets have very dark surfaces. Though initially controversial, this work was confirmed by the visit of the Giotto spacecraft to Halley's comet in 1986. Hartmann's astronomical paintings, which show the influence of the celebrated space artist Chesley Bonestell, have been displayed in Switzerland, Iceland, Moscow, and at the U.S. National Air and Space Museum in Washington, D.C. He's published several college textbooks in multiple international editions, popular astronomy books illustrated with his paintings (several in collaboration with space artist Ron Miller), a study of Spanish exploration across much of the United States in the 1500s, and two novels. He received the first Carl Sagan Medal for communicating scholarly work to the public from the American Astronomical Society, a medal from the European Geophysical Union, the G. K. Gilbert Award of the Planetary Geology Division of the Geological Society of America, and many other awards.

William Leatherbarrow is professor emeritus at the University of Sheffield, England, where he specialized in Russian studies and served as head of Modern Languages and dean of the Faculty of Arts before his retirement in 2007. He is also a lifelong amateur astronomer, with a particular interest in lunar and planetary science, who still observes regularly from his backyard observatory. He was director of the Lunar Section of the British Astronomical Association from 2009 to 2021 and served as president of that organization from 2011 to 2013. In 2020 he was awarded the Walter Goodacre Medal, the British Astronomical Association's most senior award. He is a fellow of the Royal Astronomical Society, author of *The Moon* (London, 2018), and a regular contributor to *Astronomy Now* and the *Journal of the British Astronomical Association*. Asteroid 95852 Leatherbarrow has been named in his honor, and he has also been awarded an Honorary DSc degree by the University of Sheffield for his contributions to amateur astronomy.

Baerbel Koesters Lucchitta is a scientist emerita at the USGS Astrogeology Science Center in Flagstaff, Arizona. After surviving a difficult childhood in ravaged World War II Germany, she and her husband Ivo received their PhDs in geology at Pennsylvania State University, and in 1966 they joined the Astrogeologic Studies branch, newly founded by Eugene M. Shoemaker. There she helped train the Apollo astronauts and prepared geological maps of the Moon, including a map for the Apollo 17 landing site. In addition to the Moon, she has been involved in mapping Mars and notably led the Galilean Satellite Geologic Mapping Program, which (in collaboration with Larry Soderblom) produced the first geological maps of Europa and Ganymede. She was one of the first to infer the existence of liquid water oceans beneath the ice on those intriguing worlds. During the seventies and early eighties, she also identified ice-related features located outside the polar areas of Mars, whose existence has since been abundantly confirmed. From 1986 to 1991 she served as associate branch chief of astrogeology. She is a recipient of the G. K. Gilbert Award of the Planetary Geology Division of the Geological Society of America and the Meritorious Service Award of the U.S. Department of the Interior. Asteroid 4569 Baerbel and Lucchitta Glacier in Antarctica, which she had studied carefully from Landsat data, have been named in her honor.

Yvonne Jean Pendleton is an astrophysicist with a PhD from the University of California, Santa Cruz (UCSC), master's degrees in astrophysics (UCSC) and engineering science (Stanford University), and a bachelor of aerospace engineering degree from the Georgia Institute of Technology. Hired by NASA Ames Research Center in 1979, she has used infrared spectroscopic techniques to observe interstellar dust and ices with ground-based, airborne, and space observatories. She is most interested in the materials inherited by our solar system and exoplanetary systems from interstellar disks, an interest born from an early discovery of the similarity of organic molecular signatures found in interstellar space dust and the Murchison meteorite. She retired from NASA in 2021 but continues to work with international teams using data from the James Webb Space Telescope (JWST) to further advance the understanding of our interstellar inheritance. Yvonne held several senior-level leadership positions throughout NASA during her forty-two years of service, and her accomplishments have been recognized by the NASA Outstanding Leadership Medal, the Presidential Rank Award, and several NASA Ames Honor Awards. Mentorship of early career scientists remains one of her passions, which began with numerous education-outreach efforts she led or participated in throughout her years at NASA. In addition to her ongoing astronomical research with the JWST, Yvonne enjoys scuba diving around the world and, as such, is an avid explorer of both inner and outer space.

Peter H. Schultz received his BA from Carleton College (Northfield, MN) with a major in astronomy, and his PhD from the University of Texas at Austin. His thesis, combining his interest in astronomy and geology, was published as a book, *Moon Morphology* (University of Texas Press, 1976). Following postdoctoral research at NASA Ames Research Center, he became a staff scientist at the Lunar and Planetary Institute in Houston before joining the Geology Department at Brown University. His main research interests have centered on processes shaping planetary surfaces—especially impact-cratering processes as revealed by laboratory experiments, planetary surface records, and terrestrial "ground truth." He participated in the Magellan, Deep Impact, Stardust-NExT, EPOXI, and LCROSS planetary missions, has authored over 180 research papers, and has been awarded the Distinguished Scientist Award from the Hypervelocity Impact Society, the Barringer Medal for his work on impact cratering, and the G. K. Gilbert Award from the Geological Society of America. Asteroid 16952 PeteSchultz is named for him.

William Sheehan was keenly interested in the Moon and planets as an amateur astronomer from an early age, and at one time he toyed with the idea of pursuing a professional career in science. Entering the workforce during the Great Recession of the early eighties, however, he found that the prospects for employment in astronomy beyond teaching in a community college led him in other directions. He completed an MD degree at the University of Minnesota in 1987, followed by a psychiatric residency, then spent thirty years working as a psychiatrist. At the same time, he has remained active as an amateur astronomer and avocational historian of science, with over twenty books to his credit, including *Planets and Perception* (University of Arizona Press, 1988); *Worlds in the Sky* (University of Arizona Press, 1992); *The Immortal Fire Within: The Life and Work of Edward Emerson Barnard* (Cambridge University Press, 1995); *The Planet Mars* (University of Arizona Press, 1996); *In Search of Planet Vulcan*, with Richard Baum (Plenum, 1997); *Epic Moon*, with Thomas A. Dobbins (Willmann-Bell, 2001); *Transits of Ve-*

nus, with John E. Westfall (Prometheus, 2004); *Celestial Shadows: Eclipses, Transits, and Occultations*, with John Westfall (Springer, 2015); *Camille Flammarion's The Planet Mars*, with Patrick Moore (Springer, 2015); *Galactic Encounters*, with Christopher J. Conselice (Springer, 2015); *Discovering Pluto*, with Dale P. Cruikshank (University of Arizona Press, 2018); and *Discovering Mars*, with Jim Bell (University of Arizona Press, 2021). He was editor-in-chief and contributor to *Neptune: From Grand Discovery to a World Revealed* (Springer, 2021) and has authored and co-authored monographs on Mercury, Venus, Jupiter, and Saturn for the Reaktion Books Kosmos series. He is a 2001 fellow of the John Simon Guggenheim Memorial Foundation, and a recipient of the Gold Medal of the Oriental Astronomical Association. Now retired from psychiatry, he lives in Flagstaff, Arizona.

Paolo Tanga, an Italian astronomer in France, though too young to retain any memory of the first step of man on the Moon, became keenly interested in the natural world from an early age. In response to Victor Hugo's famous question, "Between the microscope and the telescope, which of the two has the grander view?," Paolo's first love was the microscope. At age fifteen, however, he discovered the delights of the telescope and became an avid amateur observer of the planets and never went back. He went on to obtain a PhD in physics in Torino, Italy, and after a period of studying the transport of solid particles in the protoplanetary nebula by numerical simulations, he became interested in asteroids and their physical and dynamical properties. A notable achievement was successfully leading an expedition to Flagstaff at the transit of Venus in June 2012 to make coronagraph observations of the twilight arc of Venus, as part of the worldwide Venus Twilight Experiment team. Currently a planetary astronomer at the Observatoire de la Côte d'Azur in France, he is responsible for processing solar system data collected by the Gaia mission of the European Space Agency. He served as co-editor (with Marco Falorni) of *Osservre il Planeti* (Milano: di Astronomia, 1994). He has been named Chevalier de l'ordre des palmes académiques by the French Government (2015) and received with his Gaia colleagues the Lancelot M. Berkeley—New York Community Trust Prize for Meritorious Work in Astronomy (2023). Asteroid 9715 Paolotanga is named for him.

Charles A. Wood grew up reading science fiction and watching the night skies and TV news of spacecraft launches, failures, and discoveries during the early days of space exploration; most of his life since then has been exploring cosmic worlds, including Earth. Chuck was lucky to live in East Africa as a Peace Corps volunteer, setting off a lifetime of travel to sample volcanoes, rift valleys, museums, bookshops, cultures, and local foods in thirty-seven countries. He has gotten research grants, written papers and books, been on space mission teams, and given talks at conferences as all space scientists do. Since he and his wife bought the oldest house in Wheeling, West Virginia, he has renewed his interest in history and has written mystery novels set in 1850s Wheeling with an astronomer protagonist. He also wrote a fictionally enhanced biography of his 1831 house, its builder, and those times. He still works full time at a university, runs a used bookstore, drives a red sports car, and is writing his fourth book about the Moon.

INDEX

Note: page numbers in *italics* denote illustrations.